Observations Concerning the Planet Venus

Springer
London
Berlin
Heidelberg
New York
Barcelona
Budapest
Hong Kong
Milan
Paris
Santa Clara
Singapore
Tokyo

Francesco Bianchini

Observations Concerning the Planet Venus

Translated by Sally Beaumont,
assisted by Peter Fay

Sally Beaumont
The Nab, Oakthwaite Road, Windermere LA23 2BD, UK

Peter Fay
116 Lowfield Road, Caversham Park Village, Reading RG4 6PB, UK

ISBN 3-540-19980-2 Springer-Verlag Berlin Heidelberg New York

British Library Cataloguing in Publication Data
A catalogue record for this book is available from the British Library

Apart from any fair dealing for the purposes of research or private study, or criticism or review, as permitted under the Copyright, Designs and Patents Act 1988, this publication may only be reproduced, stored or transmitted, in any form or by any means, with the prior permission in writing of the publishers, or in the case of reprographic reproduction in accordance with the terms of licences issued by the Copyright Licensing Agency. Enquiries concerning reproduction outside those terms should be sent to the publishers.

© Springer-Verlag London Limited 1996
Printed in Great Britain

The publisher makes no representation, express or implied, with regard to the accuracy of the information contained in this book and cannot accept any legal responsibility or liability for any errors or omissions that may be made.

Typeset by T&A Typesetting Services, Rochdale
Printed by Henry Ling Ltd., Dorchester, Dorset. Bound by Green Street Bindery, Oxford
34/3830-543210 Printed on acid-free paper

NEW PHENOMENA
OF
HESPERUS AND PHOSPHORUS
OR RATHER
OBSERVATIONS CONCERNING
THE PLANET VENUS

From these are collected:
1. A description of its marks or Celidography.
2. Its rotation about its own axis, or Perieilesis, in the space of $24^{1}/_{3}$ days.
3. The parallelism of its axis in its 8 month orbit around the Sun.
4. Also the amount of its parallax ascertained by the method of Cassini.

NOW FOR THE FIRST TIME PUBLISHED UNDER THE PATRONAGE OF HIS HOLY MAJESTY

JOHN V

KING OF LUSITANIA, ALGARVE ETC.
BY FRANCESCO BIANCHINI OF VERONA

Domestic Prelate to His Holiness the Pope

AT ROME AT THE PREMISES OF GIOVANNI MARIA SALVIONI

Vatican Typographer in the University 'La Sapienza'

1728

WITH THE APPROVAL OF HIS SUPERIORS

Translated by Mrs Sally Beaumont, assisted by Peter Fay at the request of Dr Patrick Moore.

FOREWORD

Years ago I acquired a copy of the first book ever written about the planet Venus. It was the work of Francesco Bianchini, and it was published, in Latin, more than two and a half centuries ago. Bianchini believed that he had seen lands and seas on Venus; of course this was not so – but all the same, his book is a superb piece of history.

I suggested to Sally Beaumont that she might undertake a translation. She was well equipped to do so, and she has produced something which is the reverse of a dry historical document. It is a fascinating story, and one can easily transport oneself back in imagination to the time when Bianchini was setting up his strange, long-focus telescopes in the Vatican. With expert help from Peter Fay, I think it will be agreed that even if Bianchini has waited so long for a translator, he has at last been well served.

Patrick Moore

Students of the History of Science will find that Bianchini reveals some fascinatingly new aspects of the aftermath of the "Galileo controversy" – I found them surprising and not a little amusing.

P.F.

CONTENTS

TRANSLATOR'S PREFACE 4

DEDICATION 6

PERMISSION TO PRINT IS GRANTED   The "Imprimatur" 13

CHAPTER I   Four newly discovered phenomena 15

CHAPTER II   Markings observed on Venus 20

CHAPTER III   Construction of a globe and planisphere 49

CHAPTER IV   Description of the markings, with names 76

CHAPTER V   Rotation of Venus 105

CHAPTER VI   The parallelism of its axis 118

CHAPTER VII   The parallax of Venus 123

CHAPTER VIII   Advice on further observations 142

LETTER   From Rev. Melchior A Briga 149

THE AERIAL TELESCOPE   An investigation 160

INDEX OF CHIEF TOPICS 171

TRANSLATOR'S PREFACE

It was Venus which first awakened my childish curiosity to the wonder and beauty of the heavens. Looking back nearly half a century, I see the bare stubble of the wintry cornfield, the line of trees with one crooked poplar, the glowing lights welcoming home the returning schoolgirl, and etched forever on my memory a speck of silver playing hide-and-seek with puffy pink clouds on an icy blue sky. It was almost a mystical moment of revelation. Time and again I crept up to the bedroom window; brighter and brighter grew its steady light like a glorious lamp against the deepening blue. This was no ordinary star - was the Star of Bethlehem like this? Questions were asked, encyclopaedias consulted, and a lifelong interest was born. Now, after a lifetime of joyous sightings of Hesperus and Phosphorus in its many naked-eye and telescopic guises, it seemed entirely appropriate and almost providential that, when Patrick Moore put a request in a B.A.A. Newsletter for someone to translate Bianchini's book, I should accept the daunting yet fascinating task.

I soon recognised in Bianchini a kindred spirit whose enthusiasm shines, like Venus itself, across the centuries. His fierce determination to overcome the difficulties presented by the awkward instruments of the day, to find leisure and suitable locations for his elaborate and cumbersome experiments, to appease the church authorities who employed him and enlist the help of individual patrons, is counterbalanced by his warm humanity and disarming honesty, touched with humour and an endearing desire to reward his patrons and heroes with permanent memorials on the map of his chosen planet.

As I translated, it seemed aptly relevant but also rather sad and almost ironic that the American space-probe Magellan was daily revealing more and more of the true Venusian surface by radar, so different from what Bianchini thought. Yet though he was wrong in many of his conclusions, his painstaking methods, his use of planispheres and armillary machines to demonstrate the phases, his application of his observed movements of markings to a determination of the rotation and tilt of the axis, his measurements of parallax and his plotting of the markings on charts and globes, are all sound practice, relevant to planets like Mars with real surface markings, and very praiseworthy achievements when viewed in their historical context.

We felt it appropriate to keep the flavour of the original by retaining old names and terms, with footnotes where necessary. Bianchini has a pleasing classical style, difficult to reproduce in modern English, but its forceful and persuasive quality is reminiscent of Percival Lowell's book *Mars*, as is its subject matter so confidently expounded yet so sadly wrong. There is also a parallel in that both men had convinced themselves that they had seen permanent markings and artificial waterways respectively, and were thus predisposed to 'see' what they expected to see. As W. Sheehan says on page 83 of his *Planets and Perception* in connection with a similar case, Antoniadi's map of Mercury which he thought supported Schiaparelli's theory of a captured rotation of 88 days, "One could indulge in a certain degree of postselective discretion in interpreting the observations to fit one's scheme." Yet by chance I was observing Venus for the B.A.A. Terrestrial Planets Section, while actively engaged on the translation, and found that my drawings of extended terminator shading often closely resembled Bianchini's 'seas'. I leave the reader to decide whether Bianchini's markings were cloud features or merely faults in his telescope. I would like to believe the former, just as I would like to believe that the Martian 'canals' had some basis in reality. Indeed Lowell and other observers have seen lines and streaks on Venus that may possibly be identified with wind-flow patterns revealed in ultra-violet photographs as being caused by the 4-day rotation of the upper atmosphere. So I would suggest that perhaps Bianchini saw some cloud features which at the time had a relative degree of stability, despite my colleague's conclusions about the aerial telescope on page 164.

In conclusion I must thank the B.A.A. whose publications made this liaison possible, Mr. Peter Fay of Reading for lightening my load by translating the Dedication and Final Letter, and providing notes and help with various problems encountered, and last but by no means least Dr. Patrick Moore, who inspired the whole undertaking and was unceasingly encouraging and appreciative throughout. On one occasion he said "I am sure Bianchini would be glad," and I would like to think so too.

Sally Beaumont

DEDICATION
Translated by Peter Fay
TO JOHN V,
most serene and potent King of LUSITANIA, ALGARVE &c
Francesco Bianchini sends greetings

Your majesty received from your royal ancestors the glory of Lusitanian power so far carried to the extremities of the East and the West that the bounds of the Earth, everywhere bearing the imprint of their conquests and government, clearly show that it can hardly be advanced any further. Such is the lofty nature of your noble intentions, illustrious king, that you have not lost courage when competing with such impressive forebears in searching out fresh ways to enhance the renown of Lusitania. For when you behold the expanses of the oceans and the lands now shrunk as the result of the recent expeditions, after the renown of your grandsire has added to your possessions the harbours and continents that the sun shines on when rising and when setting, you set your heart on opening up a market for all types of knowledge in your realms, this being for the purpose of holding sway not just over the general populace but over outstanding individuals too, and learned ones at that.

You set up academies filled with the most select of minds. You fund and endow libraries, cinceliarchs and astronomical observatories. Not only do you summon professors of every discipline to the citadel of your royal metropolis, so that it will, if I may say so, tower above mere human pinnacles, but you also proceed to send your countrymen out to colonise wherever you see the finer arts flourishing, so that shortly afterwards they transfer their own outstanding discoveries, just as those of foreign minds, to the provinces of Lusitania and enrich their native soil with the perennial offshoots of their learning. You have found a means of extending the map of the world and opening the way for a new kind of triumphal glory, and a bloodless one at that.

Nevertheless, unbeaten defender of your realm and outstanding protector of the Christian Republic, you will never neglect the example handed down by your ancestors of commendation in war and of gallantry in expeditions, as we have recently learnt by the generous proof of your royal assistance in the Corcyrans' hour of peril and by removing Italy from the tyranny of the barbarians. So, keeping intact the reputation for martial courage acquired from your predecessors along with your royal inborn talents and the upbringing you received, you found the means whereby to enhance your military renown with the arts of peacetime and to extend the good name of your government beyond the boundaries of your father's triumphs.

This wonderful brilliance added by Yourself to the majesty with which you are resplendent so transfixes and captivates the eyes fortunate enough to be turned towards it, especially the eyes of those you honour with your royal gaze, and patronage, on the sole ground that they cultivated the sciences, so that vying with one another they attempt to dedicate themselves humbly to obedience to the same person and to produce some kind of memorial of their esteem for him. I indeed am scarcely allowed to be reckoned amongst this number unless my experiences of your royal appreciation and kindness had granted it, and there is not only the desire but the added necessity to try my hand at something in line with your manner of establishing advances in the Sciences under the protection of your name. Since to be sure you, my Maecenas, or rather my Augustus, contributed your assistance and patronage with regal and dignified generosity to my Astronomical endeavours, not only am I bound by the general good name of learning but also by the particular rule of the studies which I pursue earnestly to request of your kindness the opportunity of making a personal dedication. A very suitable occasion is at hand to respond to this line of reasoning: the occasion clearly demonstrates the need to search beyond the bounds of the Earth where the renown of John V, generated by his patronage of the Sciences, is marked out for immortality, and so fittingly I seem about to choose a new abode in the most glittering of all the planets which are borne around the Sun and to it both twilights have given the name Evening and Morning Star, but the people of Greece have conferred on it the title of Heavenly Venus.

It fell to my lot to detect phenomena not previously observed by astronomers in a globe of this kind, namely to map its entire surface with a remarkable variety of markings in places rendered visible to us by both stronger and weaker reflection of the Sun's light or by those rather large spots, entirely worth collating, which on the Moon we call oceans, located on that planet by the Divine Architect so as to provide us with evidence of its rotation about its own axis and indeed to make it obvious that in its own eight month's journey in orbit around the Sun the axis remains constantly pointing in the same direction, and leading us towards a fuller understanding of the entire planetary system.

I shall therefore dedicate all of these observations to your Majesty's most auspicious name, so that those alive at present and posterity should know that just as these discoveries of mine, whatever they are worth, were made under such considerable patronage, so also much more outstanding advances in the Sciences, made by others, will come forth daily if they are fortunate enough to pursue the protection of such a Patron. Accordingly, because Galileo distinguished himself by spotting the four little stars orbiting Jupiter, named for the Medici, and Cassini attained fame for the five moons of Saturn he discovered when in the employ of Louis XIV, I might be allowed to attempt this in the case of the star of the morning and evening, recently observed: namely, that I should set up a lasting monument, so to speak, of your magnificent patronage, most serene monarch; from it all will understand that I shall indeed be unequal to the task of making known the merits of your royal benevolence, yet not forgetful nor slow in substituting a more worthy herald of your glory, in other words the planet which closely accompanies the Sun and along with it scans the East and the West in the same manner. The Planet may be able to bear witness to the peoples on either side who obey your rule that their welfare is the subject of your watchful care, not only through the beneficial administration of justice, security and peace, but also through the cultivation of all branches of learning and their advancement. Long may the most foreseeing care of the Good and Mighty GOD keep you safe and flourishing, for He has raised up such a pillar of support for the prosperity of your realms, for the glory of the Christian Republic, for the defence and enhancement of the Sciences and for the protection of even the humblest of Your Majesty's servants. Amongst these you condescend to admit me and do not cease from piling on fresh reasons for gratitude as I place myself humbly at the foot of your royal throne and pen, with your permission, as a small token of my esteem these short six lines of dedication:

> Quotquot ab occiduis populos videt Hesperus oris,
> Quotquot ab Eois Phosphorus irradiat
> Concelebrant quae augmenta Tuum, REX INCLYTE, praestat
> Maxima Apollineis artibus auspicium.
> Imperium Oceano, famam dum terminat Astris,
> Jure huic Terra suas, huic dedit Aether opes.

Francesco Bianchini

> The Evening Star sees many peoples on
> the western shores, the Morning Star of East
> does shine for even more; in unison,
> illustrious King, they celebrate not least
> how far your reign excels in arts of peace
> to which the great Apollo lent his name.
> Both Earth and Sky their aid to you increase,
> the Sea your empire bounds, the Stars your fame.

P.F.

Notes.

The last line but one has been 'borrowed' almost unaltered from Virgil's Aeneid I, 287. In this Virgil makes the goddess Venus complain to Jupiter, her father, about the way in which Aeneas, her son by Anchises, is being harassed by the spiteful goddess Juno, who disliked Trojans in general; Jupiter replies and calms her fears by foretelling the great future that as yet awaited Aeneas and the new state which he was to found, i.e. Rome; in the course of the prophecy Jupiter tells her that

" A Caesar shall be born, fated to bound his empire with the sea, his glory with the stars." (Sidgwick)

The Caesar was Augustus, the first Emperor, in whose presence Virgil recited the poem. In Virgil's epic the line is:

Imperium Oceano, famam qui terminet astris.

Notice that the East, the Ocean and the Ether (or upper air) are personified. Amongst the attributes of the god Apollo were his delight in music, his love for the founding of cities and establishing their civil constitutions and much else besides; of all the gods of the Greeks it was he who had by far the greatest influence upon human life, and probably for the better.

The Emperor Augustus was well known as a generous patron of the arts. It would be interesting to know just how John V reacted to the comparison made by Bianchini. Maecenas, already mentioned, was the Emperor's principal minister and also a famous patron of the poets Virgil and Horace.

Lusitania was used as another name for Portugal. The geographer Strabo mentioned a tribe called the Lusitani living north of the Tagus. When the Romans colonised the peninsula they gave the name to the parts south of the Tagus. By the end of the 15th century the Portuguese were styling themselves Lusitanians, probably without justification.

Celidography: the study of stains or spots from the Greek Κηλις

Perieilesis: a revolution or wrapping round, especially of stars, from the Greek Περιειλησις

Parallelism: pointing always in the same direction in space, from Παραλληλος meaning 'beside one another'.

Parallax: apparent angular shift against the distant background when a planet is viewed from two different places (on Earth, in this case). The term Παραλλαξις was especially used to mean the angle formed between two lines coming from a star, one to the Earth's centre and the other to a point on the horizon. It was a measure of the star's distance in terms of the radius of the Earth.

The Popes:

Clement XI 23-11-1700 to 19-3-1721 Bianchini's patron who facilitated his early observations, especially the parallax determination of 1716.

Innocent XIII 8-5-1721 to 7-3-1724

Benedict XIII 27-5-1724 to 21-2-1730 a Dominican with a strong interest in the Liturgy and Rubrics. He restored good relations between the Papacy and the Dukes of

Sardinia and Savoy, but was less fortunate with Portugal. Pope at the time of writing of this book.

John V managed to have his royal chapel and the whole of the western half of Lisbon formed into a patriarchate in recognition of his services against the Muslims in 1716 and in the following year the Portuguese fleet won the battle of Matapan; John then wanted the Papal nuncio in Lisbon to be elevated to the status of cardinal, just as those in Vienna, Madrid and Paris but the pope at that time, Clement XI, refused, as did the next, Innocent XIII and then Benedict XIII. Finally in 1725 the pope gave him a written promise but objected to the nuncio for whom the king had demanded the favour and sent another, whom the king refused: there were then two nuncios in Lisbon, one not recognised by the king and the other not recognised by the pope. Eventually relations with Rome were suspended in 1728 - the year of Bianchini's publication. Finally in 1730 the next pope, Clement XII, did make the king's candidate a cardinal and a little later conferred the same rank on the king's patriarch. It could be said that the king thought that he ought to have a status comparable with the other major monarchs of the day, e.g. Louis XIV of France, the Emperor in Vienna of the Holy Roman Empire and the king of Spain; he perhaps also resented the consistent snubbing by Rome of his ancestors, the Braganzas, and the fact that they could trace their origins back to nobility rather than royalty.

Eventually the Pope conferred on John V the title "Fidelissimus", meaning Most Faithful; this gave the Portuguese monarchy a status in some way comparable with that of the Spanish king, whose title had been "Most Catholic" or of the French Louis XIV, who was known as "Most Christian"; compare the title "Fidei Defensor", Defender of the Faith, bestowed on Henry VIII by an earlier pope.

John V reigned 1706 - 1750. He became king at the age of 17 on the sudden death of his father, Pedro II. John was well educated but as yet rather unaware of the duties of a Portuguese king. In 1708 he married the sister of the Archduke of Austria; his first son died and the king then vowed to build a large monastery if Heaven were to grant him an heir: the site was chosen at Mafra, about 20 miles to the northwest of Lisbon. His wish was granted in 1714, so work began on a showpiece intended to rival the Spanish Escorial and the French Versailles: there was a palace, a library, a basilica and a country house; it was completed in 1735 and had proved very expensive, although the gold from Brazil had helped defray the cost. After years of prospecting with little to show for it, the Brazilian explorers did eventually find gold and in 1699 in the reign of Pedro II significant amounts began to arrive in Portugal; the crown claimed one fifth of the registered gold, leaving the rest to the prospectors as an incentive to find more. In 1728 diamonds were also discovered, the production of gold by then being in the region of 3000 lb. annually. The saying "My grandfather owed and feared;, my father owed; I neither fear nor owe", attributed to John V, sums up the growing confidence of the Portuguese monarchy since the rule of Spain was rejected, and John's grandfather, the Duke of Braganza, was proclaimed John IV. His eldest son, Alphonso, proved so incapable of governing that he was replaced by the next son, who became Pedro II, far more competent and popular.

The War of the Spanish Succession broke out during the reign of Pedro II. A Bourbon, Philip V, succeeded to the Spanish throne to the delight of Louis XIV but to the annoyance of the Hapsburgs: hence an Austrian prince was put forward by them as Pretender - Carlos III. The Hapsburg claim was supported by the Dutch, the British and eventually by the Portuguese; these were all appalled at the possibility of a future merger between Spain and France, in spite of the disclaimers of the Bourbons to the effect that the two Bourbon thrones would remain separate. At first the Allies were successful: Philip was forced to abandon Madrid and Carlos was acclaimed as king there instead; on the arrival of Winter there was a lull in the fighting, the Allies withdrew and Philip returned to Madrid with his troops; Pedro died at this time. In 1707 the war was going badly for the Allies: an Anglo-Portuguese attack from Valencia was defeated and hopes of victory grew dim. In 1709 there was a futile raid on Badajos that was undertaken against the king's wishes and those of his advisors. In 1710 the Pretender again occupied Madrid, but had to leave because of his unpopularity, but in the following year the Emperor died and the same Pretender succeeded him. This caused the Allies to have a change of heart, since a union between the Austrian Empire and Spain was even more alarming than having a Bourbon king in France and another in Spain; negotiations began that led to the Peace of Utrecht in 1712, although the Portuguese did not make peace with Spain till 1715. However, by 1711 Portugal was short of funds and the pay of the army was 11 months in arrears.

Freed from the expenses of war, John V was able to accumulate both wealth and power. He was now able to turn his mind to architecture and the patronage of learning, in short, to the arts of peacetime. He initiated a Royal Academy of History in 1720, endowed the University of Coimbra with a splendid library and saw to it that the library and palace at Mafra were also outstanding. There were books about Portuguese language and culture published under his patronage, information was copied from documents held in Rome and legal protection was accorded to Portuguese antiquities. In Rome an Academy of Portugal was founded for Portuguese artists. An aqueduct of Free Waters supplied the street fountains of Lisbon. Various new factories were set up or existing industries improved. There were new hospitals and a course in surgery was set up in Lisbon. People were sent abroad to study economics, mathematics and astronomy.

The king was renowned for his generous gifts to a variety of people and he had no need of subsidies from the Cortes, or parliament, so these ceased to be summoned. He was generous in his endowments of the Church in Portugal and expected some consideration in return.

His Royal chapel was elevated to a Patriarchate in 1716 in recognition of his crusading against the Muslims. It happened like this: the Turks seized Morea, the Emperor Charles VI declared war and the Pope appealed to Spain and Portugal to resume their crusading roles; but Spain had not yet recognised the Emperor, and so could not come to his assistance; that left John, the Emperor's brother in law, to send some ships to Corfu, previously called Corcyra (as in the Dedicatory Letter) but when they arrived they found that the Turks had left, so the fleet returned home. A fresh appeal to the Portuguese led to the victory at Cape Matapan in 1716.

John was wealthy enough to be able to stand aloof from the day to day politics of the rest of Europe. In 1728 a double marriage united the Bourbons of Spain to the Braganzans of Portugal. The only incident to cause a ripple was when some Spanish troops pursuing a fugitive violated the Portuguese Embassy in Madrid: reprisals were taken in Lisbon and British help sought and obtained, but in 1737 the situation was resolved peacefully.

John was taken ill with dropsy of the chest in 1742 and died in 1750. During these latter years the country began to fall into a state of neglect, the Queen had assumed the Regency and the conduct of affairs was in the hands of various churchmen.

Some of the Portuguese outposts had been lost to the Dutch and others. The rulers in Lisbon felt that hanging on to Brazil was of prime importance and even there the Dutch had made inroads. It appears that John IV did actually do something about both the eastern and western overseas possessions.

Christian Republic – usually this meant "Venice", but possibly here it means Christian States in general, or "Christendom".

"A personal dedication" etc. – here Bianchini implies that, unlike other Scientists, Astronomers who discover new features etc. can also name them after whoever they choose.

The Heavenly Venus – There seems to be some confusion here, since the Greek deity was Aphrodite. The Romans later identified her with their own goddess Venus. The Greeks in fact used the names Hesperus and Phosphorus (Evening and Morning) for the planet; Pythagoras of Samos is reputed to have shown that these were one and the same body.

Cinceliarch – a syncellus had originally been someone who shared a cell with an important cleric (to ensure he behaved himself), then it meant an attendant who usually succeeded to the post he was shadowing and later on it just meant an important church dignitary. There was more than one spelling. Presumably cinceliarch was derived from the same word.

P.F.

PERMISSION TO PRINT IS GRANTED

If it meets the approval of the most Reverend Father,
Master of St. Paul the Apostle's

N. Auxiliary Bishop of Boja.

RECOMMENDATIONS

"New phenomena of the evening and morning star": that is to say new phenomena of the planet Venus now seen for the first time by the most Illustrious and Reverend President. D. Francesco Bianchini. These were not to my knowledge, known to any of the astronomers and I have read through them at the bidding of the most Rev. Fr. Master of St. Paul's and have done so to the best of my abilities. This however I find most surprising of all, that if you were to ask, and I should very much like to tell you, whether his method of daily observations was accurate and 'Macrotymia' was assiduous in recollecting them, or whether indeed he showed outstanding skill in his astronomical reasoning and calculations, whether they had to be started from scratch or proved by comparison, to be honest I do not know. I do know for certain that whoever evaluates with care this most learned treatise will readily be led to conclude that this outstanding man has devoted all his mental powers to one occupation, astronomy, and has bestowed a good deal of time on observing the heavenly bodies; but when his studies have made this same man famous to the learned world in Italy and beyond the Alps, studies of other sciences and fine arts, and chiefly studies of sacred and profane history under many titles, then Rome will commend his sense of duty, praiseworthy as proclaimed by more than one, in constantly attending to his ecclesiastical functions. Meanwhile let the reader be the judge as to how profit in matters useful and advantageous may ensue from the publication of this most excellent work, that has long been awaited by learned society, not merely for astronomy but also for physiology and cosmography. This one thing I shall never cease to beg of the new Lynceus of our time: that he should not keep us waiting for the many outstanding writings, both of astronomy and of ancient learning, lying hidden in his own study, but by bringing them to light he should more and more show to foreign civilised nations that even in this present age of ours Italy is not prejudiced against outstanding men. Wherefore indeed everything in this book is in agreement with Catholic belief and sound morality and shows the dutiful intellect of its most illustrious author. I therefore give my opinion that it be entrusted to the public press, if the above-mentioned most Rev. Master of St. Paul the Apostle's will see fit.

Written in the monastery of St. Boniface and St. Alexis outside the City, 1st July 1728.

D. Didacus de Revillas, Abbot of St. Jerome's and public professor of Mathematics at Rome University.

When at the bidding of the most Rev. Fr. John Zuanelli, Master of St. Paul's, I read the book written by the illustrious and most Rev. President Francesco Bianchini about the system of Venus and its three phenomena recently happily discovered by the same, and I came upon nothing in it opposed to Christian belief and sound morality, I therefore judge it most worthy of being published in print. This is especially so when the most erudite and, in matters of astronomy, most expert author prudently put forward all his facts the more truly to be comprehended, so that no one could seek arguments from them in support one of the two most widely held systems of the world rather than of the other, and he shows clearly how the phenomena can equally well be explained according to either system. Whence astronomers, of whose discipline the Church stands greatly in need for computing Catholic Feastdays in the decrees of the Sacred Councils, will derive more concerning these same benefits and others too, and will rejoice that heaven appears day by day more and more revealed, very much as though the earth itself almost lay open, and will acknowledge that all the heavens tell forth the glory of God, when their wondrous motions and the causes of these motions have been more clearly investigated.

Written at Rome on this 24th day of June 1728 from the College of St. Clement.

D. John Francis Baldini C.R. of the Congregation of Somascha

PERMISSION TO PRINT IS GRANTED.

Fr. John Benedict Zuanelli, Order of Preachers, Master of St. Paul the Apostle's.

Notes.

*Lynceus: one of the Argonauts renowned for his sharp sight; also a nickname given to Galileo, the implication here being that Bianchini is, as it were, a second Galileo.

Systems of the world: the Earth-centred system of Aristotle in its version as modified by Tycho Brahe and the Sun-centred system (the "Solar System") of Copernicus.

D. is short for Dom, equivalent to the Spanish Don, from Dominus, the Latin for Master, c.f. Mr. 'Mister' **P.F.**

N.B. Four individuals were involved in the vetting process; the Bishop, the Master of St. Pauls and two academics (referees) who wrote "recommendations".

NEW PHENOMENA OF HESPERUS AND PHOSPHORUS

Bianchini

Translated by Sally Beaumont

CHAPTER I

Concerning three phenomena detected on the planet Venus in the year 1726, not previously observed: also concerning the order in which I propose to interpret and explain them: a fourth phenomenon also added concerning its parallax, which was most accurately investigated in the year 1716.

THE DIVISION OF THE CHAPTER INTO SECTIONS.

I. Outstanding phenomena of the heavenly bodies are reviewed, which were revealed to modern astronomers after the invention of the telescope, but were unknown to the ancients.

II. To these must be added three phenomena concerning the planet Venus, first seen in 1726.

III. Also a fourth phenomenon investigated ten years ago, namely the parallax of that same planet, from observations in 1716 of greater accuracy.

IV. The necessity to integrate the deferred publication of this fourth discovery with those three most recent observations, which are as follows: (1) The delineation of all those markings similar to those on the Moon's globe which astronomers call 'maria' ("seas"), which have now been seen for the first time all over the surface of the planet Venus, the mapping of which we shall call 'Celidography'; (2) The spinning or rotation of that same globe of Venus around its own axis in a period of 24 days which we shall call 'Perieilesis'; (3) The axis of that same planet's rotation, which constantly keeps a fixed direction in the whole of its eight-month orbit round the Sun, which we call 'Parallelism of the Axis; (4) The Parallax, or the size of the angle which half the diameter of the terrestrial globe subtends from the globe of Venus; (5) The order of dealing with these items in this short work.

I. About a hundred years have elapsed since the renowned invention of the Telescope at the start of the last century, during which so many secrets of the heavens have been uncovered, that all the wealth amassed from that source for the enrichment of Astronomy might seem to be almost exhausted, most particularly that sought in connection with the Solar System. For Galileo, like a true Lynceus, detected spots on each luminary, and from the solar ones discovered the rotation of that globe around its own axis; around Jupiter he discovered four satellites and their orbits; on Venus he noted phases, similar to those of the Moon. Huygens discovered Saturn's ring and one satellite; Cassini found four other satellites besides that of Huygens. Gassendi was the first to announce a transit of Mercury across the Sun, as Horrocks did of Venus. Others revealed that comets originate beyond the Moon, or even (now it is permitted to say so) beyond Saturn. Some showed the number of fixed stars immensely increased: others saw streaks on Jupiter, and its rotation round its own axis; nor was Mars free from reports of blemishes. It seems that there was nothing now remaining which Nature had reserved for the diligent research of ourselves or our descendants.

II. But who could ever say that the wealth of the Heavens was exhausted, surpassing as it does the measure of our senses and imagination in number and immensity? On the globe of Venus alone, a short period of the year just past has provided me with three phenomena to observe, unknown to earlier times; ten years previously a fourth had been revealed. Each one of these would easily be considered by devotees of astronomical studies no small addition to the inheritance bequeathed by our ancestors, to the promotion of science and, what is more, to Cosmography and universal Physics, if a more skilful observer of sights like these were blessed with a favourable opportunity and the diligence to use it.

III. I had reached this conclusion even then, ten years ago, when that opportunity for observing was first presented to me. That was the Parallax of Venus, which could be measured very accurately because of the proximity of Regulus, if I could watch this star a long time in the same field of view of the Telescope with the Planet as they approached the meridian in conjunction: and this I fortunately achieved on July 3rd, 1716, with a 23-palm[1] telescope made by Guiseppi Campani. For on this depends not only our knowledge of the distance of the planet Venus from the Earth, but also our distance from the Sun, and a more exact measurement of the whole Solar System, which I frankly doubt we can achieve as accurately by any other means.

At first I was gripped by a desire to make public that observation, particularly when several of my friends urged me, who knew it had never been attempted before. I put it off, however, as after an interval of eight years the opportunity of repeating the same observation on almost the same day of the year would recur, so that I could confirm my discoveries with a second try. But it was not possible for me to repeat it, although a completely clear sky gave a good chance of observing, since I was denied the use of the 23-palm telescope which I had employed eight years before, and the aperture of shorter telescopes[2] did not permit a large enough magnification for Regulus to be seen by day in conjunction with Venus on the meridian. So although my hopes were dashed of repeating the experiment in 1724, I thought my first results of observations of that planet ought to be suppressed no longer, but should be made known to all connected with Astronomy, so that it could

---

[1] *The Roman palm: the following information was supplied by Dr. J Landels, former Senior Lecturer in Classics at Reading University, now retired.*

There were two different measurements based on the hand. The Minor Palm of 4 'digits' amounted to 7.4 cm. whilst the Major Palm or span of 12 'digits' equalled 22.2 cm. (8.74 Imperial inches to three significant figures). The latter was of great antiquity and was the one used by Bianchini.

Minor Palm   Major Palm

See Tab VIII for illustrations of Campani's telescopes, which enjoyed a very high reputation at the time. **P.F.**

² *Objective lenses of rather long focal length were the only known means of minimising the effect of chromatic aberration (spurious coloured fringes) that were inevitable with the use of a single lens. Newton had despaired of finding glass with a different refractive index from that in common use, and so advocated the use of reflecting telescopes instead - all colours are reflected equally whereas they are refracted differently as Newton's prism experiments demonstrated beyond doubt. It is ironic that two years after this book was published the English barrister, Chester More Hall, was busy designing the first achromatic doublet - a biconvex crown glass lens in contact with a concave-plane flint glass which corrected for chromatic aberration and made manageable telescope lengths possible! It was John Dollond (whose name is still around today) who manufactured such doublets in quantity. The reader will appreciate that any sort of accurate equatorial mounting was out of the question and so Bianchini just had to align his telescope correctly and leave it fixed for the duration of his experiment.* **P.F.**

be repeated more accurately by those with greater skill when after another eight years Venus returned to the same position with respect to the Sun and the Earth, and Regulus could next be seen on the meridian on July 4th, 1732.

IV. So that I did not have to publish alone this experiment which is very useful to Astronomers, Cosmographers and Physicists, a fortunate opportunity arose in 1726 which provided the means of making other observations concerning the same planet, first at its evening elongation from the Sun when it bears the name Hesperus, then in the morning one when called Phosphorus. Since both sets of observations went exactly as I hoped, devotees of these studies will receive the promised observation of Parallax supplemented with several other discoveries.

There are indeed three outstanding discoveries of that year which I have added. The first is a description of the whole globe of the planet and the markings observed on it, which we shall call in the Greek κηλιδογραφια, or Celidography. The second is the measuring of the spinning or rotation of that globe, completing a full circle around its own axis in 24 days: and the fact that its axis keeps pointing in the same direction in completing its eight-month orbit round the Sun is what I call the third of the discoveries I have made this year. That revolution, or spinning and rotation round its own axis we shall call περιειλησις, or transcribed from the Greek, Perieilesis. We shall use the same language to describe the propensity of the axis to keep the same direction constantly, παραλληλισμον or Parallelism; and since it has been possible even to measure the angle which half the Earth's diameter subtends from the distance of the celestial bodies, we shall keep to the Greek language and call it παραλλαξις or Parallax. Therefore the three discoveries of this year, and the fourth one that was made ten years previously about the planet that was given the name Venus by the ancients, or Hesperus and Phosphorus, will be the subject of the present small work, and they will be dealt with in this order:

1. *Κηλιδογραφια*: Celidographia, or description of markings,
2. Περιειλησις: Perieilesis, or rotation round its own axis,
3. Παραλληλισμον: Parallelism of the axis in revolutions,
4. Παραλλαξις: Parallax, or the size of the angle which half the Earth's diameter subtends from that planet.

The histories of these four discoveries must be explained one by one, with the conclusions deduced from them or still to be deduced. Also there will be an explanation of the method used to set up each observation carefully, so that they can be repeated at will by whoever is studying this aspect of Astronomy, and whoever loves to revere and contemplate the workings of Divine Wisdom in the disposition, immensity and motions of the heavenly bodies.

Although my investigation of the Parallax was earlier than my discovery of the markings and the rotation of the planet round its own axis, I will however keep to the stated order of dealing with them. For we must first consider the globe of this planet and give a full description of its markings, which is obtained by simple observation through larger telescopes. Then from the differing positions of the markings, which move in an ordered way day by day, the planet's rotation round its own axis is seen. Then from the parallel circles described by the markings in that rotation, and by observing their meeting and intersection, at gradually varied angles near the point of the successive progress of the planet in its eight-month orbit, with the terminator which indeed divides the illuminated hemisphere from the dark one, the Parallelism or inclination of this axis of these revolutions or rotations is deduced, which remains everywhere constant. Finally, having shown which of the many parts of its eight-month orbit will give a greater chance of investigating its parallax, we proceed to achieve this through other selected observations. Therefore let us take our starting point from the Celidography of that globe, the planet Venus, in that it is the more simple observation on which the rest is based.

CHAPTER II

A DESCRIPTION OF THE MARKINGS OBSERVED ON THE PLANET VENUS
(OR MY FIRST DISCOVERY) WHICH WE CALL CELIDOGRAPHY

Summary of the Chapter

I.   *The first opportunity for observation.*

II.   *Other observations of the lunar marking called Plato, accomplished in the same period, 1725 and 1727.*

III.   *Observations of Venus undertaken in 1726.*

IV.   *Selection of the sites to perform the observations, first at Rome and then at Albano, performed with 100-palm telescopes.*

V.   *Selection of the time, and types of eyepieces.*

VI.   *The first observations undertaken in February disclose markings on the glob of Venus (called Hesperus during its evening elongation from the Sun), and their rotation round the planet's axis in 24 days, from which is revealed the Celidography of more than half of Venus' globe.*

VII.   *The circumpolar markings of that planet, which could not be described from the February and March observations as they were not lit by the Sun, were revealed from other observations made in May and June 1726 when the South Pole was presented to the Sun and to our gaze, and in July 1727 when the North Pole was presented, to complete the Celidography of the whole globe.*

VIII.   *The possibility of more exact observations and descriptions if, at times indicated here and more fully explained in the last chapter, telescopes are directed at that planet.*

IX.   *The reason for the term Celidography being used for this description.*

X.   *Precautions to be taken to give a clearer view of the markings; and concerning their likeness to the markings generally called "maria" on the Moon.*

I. I acknowledge, in obtaining the first opportunity for this discovery, the help of the most eminent and reverend Prince Melchior, Cardinal of Polignac; who in accordance with his innate mental perspicacity towards all sciences, and with that magnanimity and keeness of intellect which is his strength, in the midst of the most arduous ministerial duties of the realm in the courts of Europe, specially entrusted to him with public safety in mind, never failed to pursue those disciplines most worthy of a Prince. Thus we can most justly repeat in his case what Cicero[1] said in his Second Book of the Academic philosophy, about Cato, Scipio and Lucullus, that in the midst of their noble offices and duties relating to war and peace they never neglected Philosophy. "The qualities he showed deserving recognition by popular esteem, are almost the same in our estimation and that of others abroad. For we have experienced, along with many, these qualities he revealed abroad, and often too have shared these qualities at home with a few." Therefore the matter proceeded favourably under the auspices of that Cardinal who is most learned and most desirous of promoting the sciences. For he had been wanting for a long time to use the largest telescopes of 100 and 200 palms, made by the famous craftsman Guiseppe Campani, some time ago, for celestial observations. To him he had generously supplied 17 years previously the necessary money to construct a wooden machine and set it up, so that if the object glass were raised up high and held firmly, an eyepiece could be placed at a distance of 100 or 200 palms adjusted with a cord by the method of the famous Huygens to the correct focus, and thus very highly magnified images of the heavenly bodies to which the telescope was directed could be shown at a very wide angle[2]. While Campani was setting up a machine of this kind around August 1709, the Prince, who was then serving as President of the Court of the Holy Roman Rota[3], was asked to take on the duty of an ambassador to King Louis XIV of immortal memory, to stabilise the peace of Europe at the Council of Utrecht. So for 16 years he willingly gave up his plans for experimenting in order to serve the needs of the Christian Republic[4], "since such great qualities of virtue and genius had long been absent, lost to the eyes of both the public assembly and the Senate House". Just like Cicero's Lucullus, he returned again to Rome (and as Lucullus came with a triumph, he came with the decoration of the Sacred Purple, which his great achievements had long merited) and he did not cease to promote scientific studies even more keenly, especially astronomy. Therefore he arranged with the heirs of Campani, who was now dead, for telescopes with the prepared machines to be set up to look at the heavens.

---

[1] *Cicero: Roman statesman and orator, 106-43 B.C., who praised the famous Roman politicians and generals here mentioned in similarly flattering terms as Bianchini applies here to his patron, Prince Melchior.* **S.B.**

[2] The 100-palm telescope: this would have been about 73 Imperial feet in length, or about 22 metres. **P.F.**

VERSIONS OF THE LONG FOCAL LENGTH "AERIAL" TELESCOPE

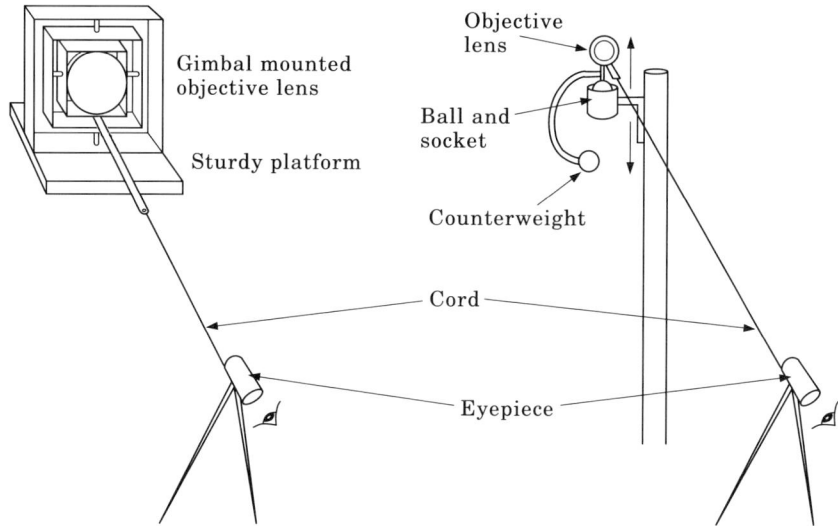

Bianchini's telescope may have looked like either of these types, or something in between.

The object glass was mounted in some type of universal joint. It was turned by the cord pulling on the rod, the whole mounted on a "sturdy platform", which could have been supported on a tall pole or it could have been the 'gimbal' version placed on a suitable wall etc. on its own, although this would severely limit the number of suitable sites available, especially for viewing bodies well above the horizon. From what follows in Section IV of this Chapter it would seem that the "wooden machine" in use here was a folding device which could be extended, rather like a large car jack, but relying on "little beams" (which might have resembled a trellis work).

A refractor is much more tolerant of slight misalignment than a reflector. The long focal length reduced the effect of chromatic aberration.

The silk cord. The telescope he used on 26th February 1726 was of 88 palms (63.8 feet or about 19.5 metres; 1 'palm' = 0.73 feet approximately). Obviously a tube of this length would have been unwieldy, to say the least, and Huygens suggested the use of a cord in the so-called "aerial telescope" to link the eyepiece and the object glass.

The eyepiece tube containing a Huygenian doublet: two plano-convex lenses, the first of focal length $4x$, the second of focal length x, separated by a distance $2x$: these are the dimensions originally proposed by Huygens, to give a wider field of view, but the 'traditional' version of this eyepiece has the first lens of focal length $3x$ and this reduces both chromatic and spherical aberrations.

To help steady the tube there were two legs that rested on the ground.

A speck of dust on the front lens of the pair (the field lens) might be mistaken for a marking on the planet: rotating the tube would settle the doubt. (A speck on the object glass would not show at all). **P.F.**

[3] Holy Roman Rota: a tribunal within the Papal Court.

[4] Christian Republic: he probably means "Christendom" i.e. the Christian States.

II. The telescope was set up on the Palatine Hill on the evening of August 16th, 1725, not without considerable success at the first attempt. Although we were only able to direct the telescope of 150 palms focal length towards the Moon that night, we saw on the marking known as Plato a phenomenon not previously observed. The Moon's phase was just a little past First Quarter, which had occurred the day before, and Plato lay exactly on the terminator. The whole rim which surrounds that feature, like a deep pit with a very high wall on all sides, appeared brilliantly illuminated by the Sun's rays. The bottom of the pit, with no solar light reaching it, still looked very dark. But a shaft of reddish light traversed the middle of that dark region, like a bar stretching straight from one end to the other; just as the Sun's rays appear to the eyes in winter when admitted through the window of a closed room; or shafts of sunlight projecting from distant holes in the clouds; or the tails of comets streaming far through the clear night sky, as we remember seeing from the comet conspicuous throughout Europe in 1680 and 1681. A similar appearance I have never seen before in Plato or any other lunar marking, as I show in the drawing below.

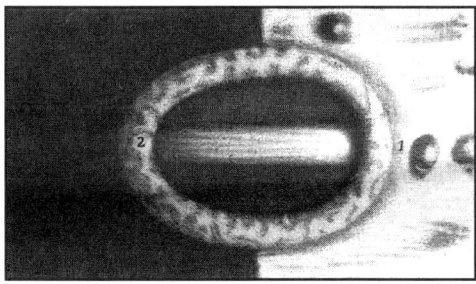

The lunar marking named Plato, and the reddish ray of sunlight stretching across its dark floor, which proceeded from the bright wall facing the Sun.

An observation made on the Palatine Hill in Rome on August 16th 1725, an hour and a half after sunset, with a 150-palm telescope made by Guiseppe Campani.

The proposition that astronomers and physicists should consider, is that they should judge whether this is an indication of the existence of a breach in the wall of Plato that faces the Sun, through which the rays pierce as they do through a window; or perhaps they prefer the view that the rays are refracted, as they are bent from the top of the wall towards the floor and become red, just as happens in our atmosphere when the Sun is rising or setting, and this might be an indication of some denser medium existing in the form of an atmosphere round the Moon's globe. We were quite encouraged that day by our first experiment of this kind, and were preparing to turn our telescopes to the other planets in a week's time, when an area of ground of sufficient scope in the direction of their orbits could be obtained to extend the cord to 200 palms[5].

---

[5] *Evidently Bianchini and his companions had a range of objective lenses of differing focal lengths and requiring different lengths of the cord. Just imagine the problems that a rigid 'tube' of 146 feet would entail!*

But my astronomical efforts were nearly thwarted by my study of Antiquity. For on the next day, August 17th, I was on the part of the Palatine Hill[6] which looks towards the church of San Gregorio Magno to the southeast on the slope of Scauri. While I was exploring the ruins of the Palace of the Caesars to investigate its layout from the surviving walls, (the discovery at that time of an extensive basilica, and the chambers belonging to it, in the Farnese gardens provided the opportunity for this investigation. The grandeur of its structure and ornamentation has revealed to the eyes of the citizens of Rome the magnificence of its ancient Emperors, an example of which I will provide, God willing, when I have worked out the ground-plan), and while I was carelessly running about to take the measurements of the surviving rooms in the east wing of Augustus' home, whose ruined walls survive in the vineyard of the English College, I fell into a broad, deep hole in the pavement which I had not noticed as I rushed around with my eyes fixed on the point to which I was going to measure; I broke my right thigh, and by God's mercy was only saved from the imminent death which threatened me by pressing with all my power with both hands and my left foot against the sides of the hole, to sustain the weight of my body and avoid falling headlong into the pit forty palms below, the depth of which I knew from my measurement of the lower chamber. The injury to the thigh interrupted the observations I had begun, which however with God's grace I was able to resume even more successfully at the start of the following year, 1726.

Another observation of lunar markings should be included in this section, made with the same telescopes on September 22nd, 1727, and repeated more than once later. We discovered on the Moon's surface a kind of incision proceeding in a straight line for a certain distance which we were able to measure. We also saw some small polygonal areas bounded by equal straight lines. So it should sometime be possible for future observers to turn their telescopes on the same objects and take careful note if any changes have taken place on the lunar surface. For as regards straight lines, any slight variation from our observations should be discernible, unless their regularity is constant.

And so the 150-palm telescope of Campani was directed at the Moon on the aforementioned day, September 22nd, 1727, just after sunset. On the orders of His Eminence the Cardinal de Polignac, it had been set up in a very extensive area where the remains survive of the walls forming the royal

---

[6] *The Palatine Hill, site of the earliest Roman settlement, where Augustus built his great imperial palace, which was enlarged and embellished by later Emperors. After the fall of the Roman Empire the Palace decayed, and was covered by gardens, convents etc. Systematic excavation did not begin until 1871, so it seems that Bianchini was a pioneer in this as well as in astronomical observation.* **S.B.**

hall or basilica of the Palace of the Caesars, discovered a little earlier and cleared of rubble, on the Palatine Hill within the Farnese Gardens, so that we might re-observe the floor of Plato, on which we had detected in 1725 the shafts of sunlight described a little earlier. But the Sun's light had not yet reached the surrounding walls of that marking; only a small part of the floor was over the terminator on the light side. The rough areas nearest that part of the floor alone were visible, and are indicated by the letter E in the diagram below. Letter A indicates the marking which Riccioli and other astronomers have named Aristoteles, and B is the one they call Eudoxus. D has not been named, but is shown exactly on the very accurate Moon map "Selenographia" printed at the Royal Press in Paris by the Academy of Sciences[7].

Lunar markings and shapes with straight lines observed at Rome with Campani's 94 and 150-palm telescopes on the eighth day of the lunation on August 23rd and September 22nd 1727.

A = Aristoteles, B = Eudoxus, C = Plato.
1–2 is the straight line leading to the small marking 3.
4 is a triangular area.

---

[7] *D is the crater now known as Cassini, 1-2 is the Alpine Valley. Hevelius observed in Danzig, and his 'Selenographia', the first proper atlas of the Moon, was published in 1647.* **S.B.**

If from that spot D a straight line is extended to E, that represents the line of the terminator that night. The small section DBAE was still in the illuminated portion while the whole of Plato still lay in the dark portion on September 22nd. Almost midway between the borders of C and D was seen the gash stretching in a straight line 1-2, like a kind of long ditch, the line of which pointed towards the small feature 3 without actually reaching it. A very clear view of this straight gash was given by the 150-palm telescope described, particularly that night when seeing was excellent. However, we saw that same ditch or straight incision even with much smaller telescopes, namely of 25 palms, in the subsequent months and on the 'Selenographia' map of the Royal Academy of Sciences it is shown lightly shaded. The best time for observing it is First Quarter, when it is exactly on the terminator, and that had occurred the day before, September 21st of that year. His Eminence the Cardinal, who noticed it before the rest, remembered seeing in the same region also a small area like a pentagonal shape, and there also I have indicated by the figure 4 another triangular figure that I noticed. On the following day, September 23rd, the whole of Plato was illuminated, and displayed on its floor a rather long shadow from the high elevation on the rim E, which projected almost to the centre of the floor, C. But no trace was seen of the shaft of sunlight which had been observed on August 16th, 1725, through the shadow, perhaps because the gap through which it penetrated in that first observation of 1725 was not reached by the Sun at that particular height that it attained on the second observation of 1727 above the plane of the marking and its floor, C.

The length of that straight incision is as great as the distance between the rims or edges of Aristoteles (A) and Eudoxus (B). Measured by micrometer it is 1/32 of the whole lunar diameter. As the lunar diameter is a measurement long known to astronomers since the invention of the telescope and the fitting of micrometers to it, namely a little over a quarter of the Earth's diameter, that is to say about 2,200 Roman miles, a thirty-second part of that sum gives the length of that straight incision as about 70 Roman miles[8]. (These are the miles we see marked on the Appian Way, the ancient measure of Vespasian and also the modern one of today.) For no one has any more doubts about the size of the Earth's diameter since the very careful measurements of the Royal Academy of Sciences published on page 247 of the book entitled De la Grandeur de la Terre, and other estimates not widely different from those made by the Royal Society in London. It is about 6,538,594 Parisian measures (Toeses), which is equivalent to 8,525 old and modern Roman miles approximately. One degree of a great circle round the Earth's globe at latitude 44°8' according to Cassini's calculation, is 57,130 Toeses, which is 76 Roman miles and 17 paces. Our experiments are in close agreement with these measurements and, God willing, will be published along with the line of Meridian which passes through Italy, from the Roman shore near Ardea leading through the city itself and the Baths of Diocletian to the Sundial of Clemente, then through the shrine of San Silvestro on Mount Soracte, through Eugubium[9] and reaching the Adriatic coast just above Rimini; or the line which runs parallel to it from the house of M.Costagutus on the shore at Anzio to the vineyard of the Irish College below Castel Gandolfo, leading on to Sabina through Vaco village and the mountain sheltering it, Monte Coscia, then through Monte Acuto, the highest point of this section of the Apennines above Cantianum on the Flaminian Way, finally reaching the eastern seaboard at the port of Rimini. Part of this line of longitude I published in the Appendix on Cartography to my 'Extracts from the History of the city of Rome' published in Italian at the Vatican Press in 1724.

But this is a digression concerning my observations of lunar markings and the measurements of the Earth's globe. We must return to the markings recently discovered on Venus.

III. The globe of Venus or Hesperus was conveniently placed for observation in the evening twilight, displaying to us a hemisphere half sunlit and half in the shade like the First Quarter Moon. Convenient too was the arrival of a nobleman named Hope from one of the foremost families of Scotland. He had long ago attracted the attention of His Eminence the Cardinal de Polignac, not only on account of his noble birth but also for his innate scientific genius. This quality he had observed in the young man during his peace mission in Holland, but he had also been recommended to him by His Serene Highness the Duke of Lotharingia, in whose Academy for the cream of European nobility he had spent a year, impressing not only with his astronomical studies but also with his knowledge of antiquity. When he learnt of our proposal to turn the 100-palm telescopes made by Campani on Venus, the use of which was available to me both at Rome and Albano, he said he would attend the observations at both places; and other eminent men were often present too. Outstanding amongst them were His Excellency the Duke of Jubenati, brother of His Eminence Cardinal Judice, who had recently acted as ambassador for the Catholic King of Spain at the court of Louis XIV. He too augmented the considerable praise he earned for the excellent performance of his ministerial duties with the renown which accrues to such great men from their patronage of the sciences and their diligence in promoting them.

---

[8] *1 Roman mile = 1620 yards (1,000 Roman 'paces')*
1 Toes = 6 Paris feet, = 2.1 yards. **P.F.**

[9] *Eugubium - an old settlement just East of Todi, 20 miles South of Perugia.* **S.B.**

IV. The interest of many spectators, therefore, made me keen to repeat the observations quite often. The opportunity to make them was not so easily acquired, however, through the lack of suitable places to set the machines up. A site in the open air had to be sought, on the west side of which the substructure could be elevated to at least 20 feet, and on it a level platform open to the sky, on which the wooden machine might be placed which was intended to hold up the object glass in such a position that the thread of the Huygenian device holding it up could be freely let down, rotated and extended to the eyepiece of the telescope situated at the required distance of 100 palms. The altitude of Venus above the horizon at that time was about 40° (as the planet was where the ecliptic descends at 90°) and this was favourable for a clearer view, in that it was free from the mists which haunt the regions nearest the horizon and render stellar images unsteady there. But to achieve such a height required quite a considerable elevation of the object glass, which the little beams of the wooden machine could not provide alone without the addition of some structure beneath.

We found two gardens in Rome suitable for such an elevation and facing that part of the sky, so that both lenses of the telescope could be set up there and directed towards the planet.

One very spacious site was discovered in the Palazzo Barberini on the Quirinal Hill, where the side facing the south is joined by a wooden bridge to the higher level of the gardens and provided with a wide ditch surrounding the palace like a fort. We went down to observe on a level area by this ditch with the eyepiece supported on its device. We put the larger machine necessary to support the object-glass on the wooden bridge already mentioned, where it could be easily raised or lowered depending on the planet's elevation.

Another site, admittedly smaller but suitable for observing at least for an hour, was obtained on the Esquiline Hill near the titular church of St. Prudentiana in the residence built and set aside for the hospitality of Dignitaries and Priests of the Syrian Church who serve their country on sacred embassies, by the illustrious and reverend Praesul Athanasius for Saphar, Bishop of Mardina. The garden there was elevated towards the west, and offered a suitable site for placing the wooden machine. The level area below, planted with a shrubbery stretching towards the public Via Sixtina from the Four Fountains to the Basilica Liberiana, afforded a facility for setting up the eyepiece and extending the cord to a distance of a hundred palms.

Outside Rome I had an equally suitable site at Albano on a raised lookout place belonging to a house opposite the Abbey of St. Paul, built by the Knight Carlo Maratti, the celebrated artist. When we raised the telescope's object-glass up to a height of 70 palms on that look-out place and set up the eyepiece on a lower site in the open air, the situation was most suitable for extending the cord to the correct distance and pursuing the observation for a whole hour from the beginning of evening twilight.

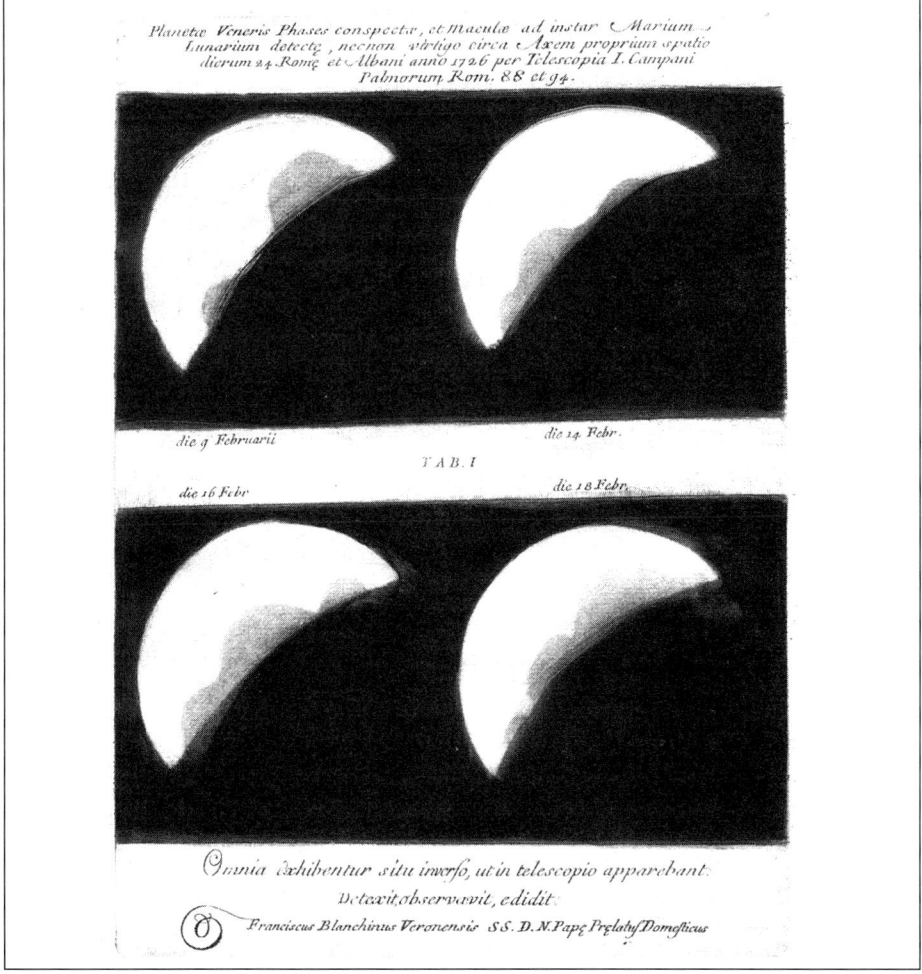

TABLE I

Phases of the planet Venus observed, and markings like the lunar maria detected, showing the rotation around its own axis in 24 days, at Rome and Albano in 1726 with telescopes of 88 and 94 Roman palms made by G. Campani.

All are shown inverted , as they appear in the telescope.

Discovered, observed and published by Francesco Bianchini of Verona, Domestic Prelate to his Holiness the Pope.

TABLE II

Phases and markings observed on the planet Venus during the evening elongation before inferior conjunction from February 9th to March 7th 1726. Inferior conjunction followed on April 6th. The phases are shown in their true orientation.

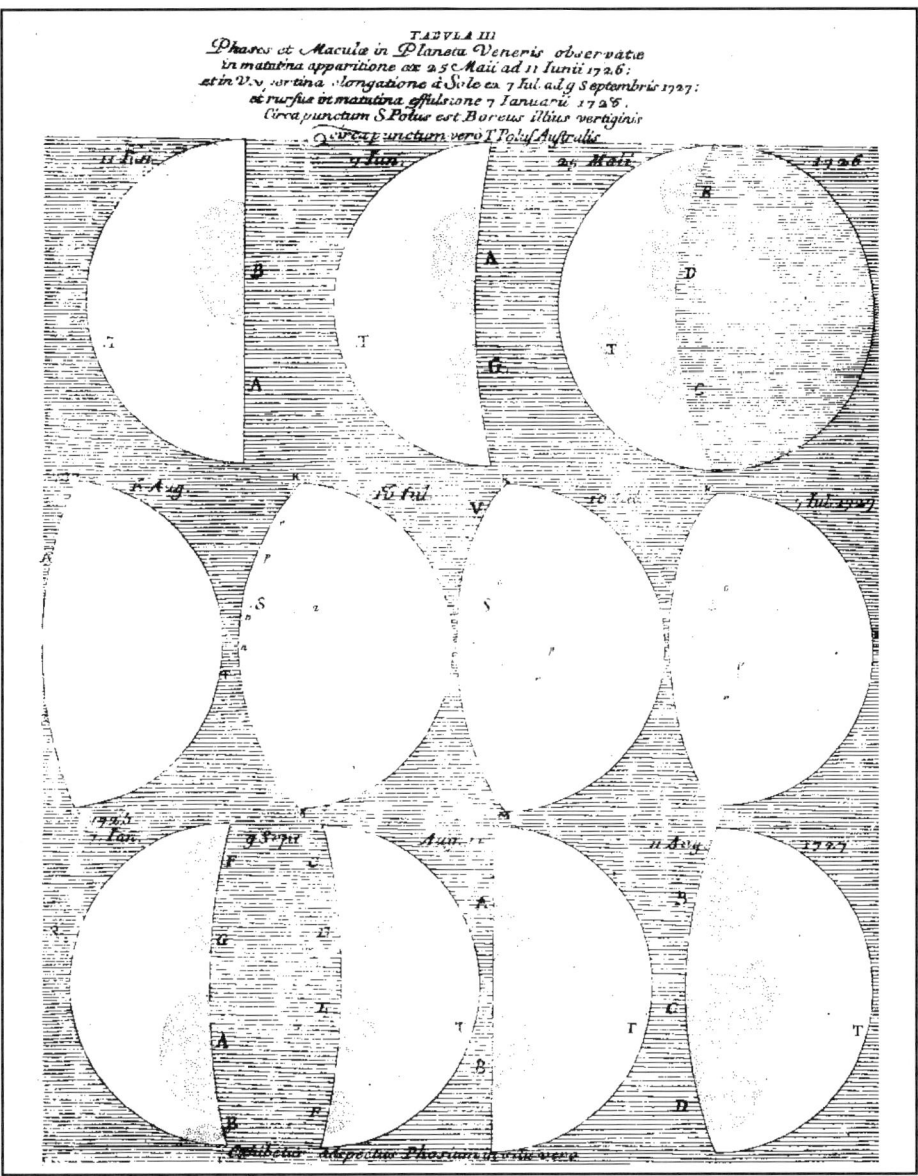

TABLE III

Phases and markings observed on the planet Venus during the morning elongation from May 25th to June 11th, 1726, and during the evening elongation from July 7th to September 9th 1727, and also in the morning elongation on January 7th, 1728. The North Pole is at the point S, the South Pole is at T.

The phases are shown in their true orientation.

V. After the selection of these observing sites, therefore, whenever we had the opportunity in February and March we made daily notes of the markings which appeared on Venus' globe. They were similar to the larger lunar ones which can be seen with the naked eye on that luminary, called 'maria' by astronomers on the 'Selenographia' map, for example Mare Crisium, Mare Serenitatis etc. They are really areas on the surface of the globe which are less efficient at reflecting the bright sunlight.

In order to see these markings more clearly it is necessary to choose days free from mist, but also to wait until twilight is more advanced, at least half an hour after sunset. For the naked eye features called 'maria' on the Moon appear to be surrounded only by a washed-out pallor if observed at sunset, but stand out more clearly half an hour later in the darker sky; and in the same way the brighter regions of Venus show in the telescope a much greater contrast with the paler parts of that disk if atmospheric glare is diminished and does not distract and dazzle the eye.

Moreover the aperture of the object-glass used to focus the rays for this observation has to be decided. Among several we tried, the best was one which had a diameter of 4 unciae[10] in a telescope of 90 or 100 palms.

Finally an eyepiece should be selected whose focal length is no more than 10 unciae and no shorter than a Roman half-palm.

VI. Having taken these precautions, I made observations from February 9th to March 10th, which I show in Tables 2 and 3, with separate drawings for each day when observations were possible.

Thus when we compared the positions of the markings seen on different days around the same time of evening twilight, it was easy to measure their daily progress, which caused them to move about fifteen degrees every day from west to east. This is obvious when we look at the drawings, in Tables 1 and 2, showing the disk of Venus from February 16–20th. For the larger marking C, whose rounded tip 3 was observed by us on February 16th to lie along the line through the centre of the disk at right angles to the terminator SCR, had moved toward the limb R on February 18th, two days later. On the 19th it had approached the limb much more closely, and almost touched it on the 20th. So within a period of 6 days the marking C was observed to have completed a quarter of its revolution, from C through B to R. And so we realized that the larger marking A (Table 2, Figure 1), which on February 9th had proceeded beyond the central line PQ, five days later would already have rotated beyond the limb R and could not be seen by us on the 14th, but only the smaller marking B which followed it (Table 2, Figure 1), which on the 9th was on the quadrant PS.

---

[10] *uncia = a twelfth part i.e. a twelfth part of a foot, or an 'inch', or a twelfth part of a palm as here.* **P.F.**

After the marking C3 followed other smaller ones, which adjoined it, D4 and E5 (Table 2, Figure 4), whose progress is shown to be similar on the same drawings. At last on March 5th, when the spots on Venus' disk had returned to almost the same positions, and the same ones A and B that were seen on February 9th were visible again (Table 2, Figure 1 and Figure 8), we discovered that a full revolution was completed in 24 days, and dividing the 360° of the whole circumference by 24 days, we found that a single day's progress was 15°. The drawings of Venus' disk and markings I show inverted in Table I, as they appear in a telescope fitted with only one lens* in the eyepiece, as is usual for observation of celestial objects. It is preferable, however, to show them thus inverted, so that they can more easily be recognised by those who want to try the experiment when the planet returns to a similar aspect towards us at stated times.

These same markings ought, however, also to be shown in their true situation, and this is provided in Tables 2 and 3, so that it will be easier to see the nature of its revolution and illumination, the position and tilt of its axis, if the power of imagination is helped by the correct orientation of the drawings of each phase.

VII. It is obvious that within the space of 24 days, the whole globe of Venus will be presented to us because of this rotation around its own axis, but there are two limiting conditions that must be fulfilled by this rotation. The first one is that the hemisphere of the planet's globe turned towards us should be uniformly lit by the Sun, in such a way that we share with the Sun a view of the whole illuminated sphere throughout the 24 days of the rotation. The second one is that the axis of rotation should coincide with the line of the terminator and each pole remain fixed relative to this. For if the globe is thus orientated, the various individual parts of its surface will gradually display themselves to our view during the course of a full revolution, while they describe their parallel circles at right angles to the axis. For thus no region or marking will be hid from us, since whatever is presented to our view in such a position will also be lit by the Sun's rays. Therefore a complete description of all the markings on the globe of Venus, which we have decided to call Celidography, could be completed within 24 days if these conditions were fulfilled.

But in actual fact neither of these conditions is granted to us by the Heavens. For in order that the same hemisphere of the planet should be presented both to us and the Sun, it is necessary for us to be placed right on a straight line drawn from the Sun through us to the planet. This can happen in the case of the three superior planets, Mars, Jupiter and Saturn, but never in

---

* *The lens combination in Huygens' eyepiece is equivalent to the "single lens" that Bianchini has in mind here.*   **P.F.**

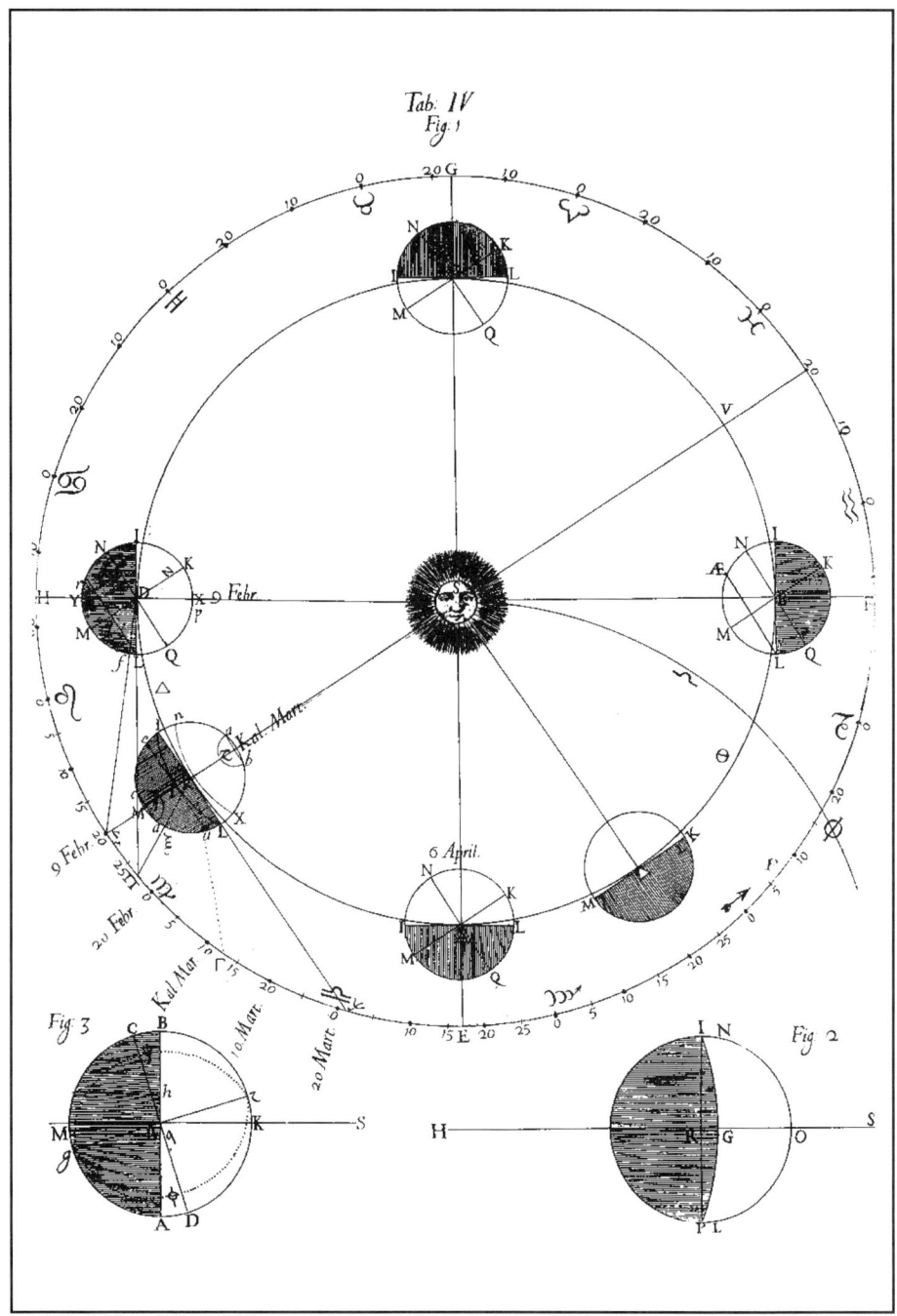

TABLE IV

the case of the two inferior planets, Venus and Mercury. And so, since the complete hemisphere of Venus facing the Sun is not at the same time presented to our sight, we must consider how it is possible during successive revolutions of the planet round its own axis for all sections of the globe to be seen and recorded by us, by waiting until each section in sequence moves round day by day. We realize this can be more easily understood from Table IV, which shows a plan representing the movement of Venus round the Sun and its appearance both from the Sun and from the Earth at the different stages of its 8-month orbit.

To understand Table IV, consider that one is looking down from above perpendicularly on the plane of the ecliptic from its north pole. Let point S represent the Sun, the centre of this plane. Line AS is the radius of the circle ABCD traversed by Venus in its orbital revolution. Admittedly it is an ellipse but very nearly a circle. If the radius AS representing the distance of Venus from the Sun is divided by three, and if this line SA is extended by one of these thirds to E, this line SE will give the average distance of the Earth from the Sun. This is one and a third of Venus' distance from the Sun, or that is to say SE to AS is in the ratio of 4 to 3. In the course of its 8-month orbit round the Sun, Venus accomplishes a quarter of its orbit, from A to B, in 56 days. After another 56 days Venus reaches the end of its second quarter at C. Then the third quarter CD is accomplished in equal time, and finally from D it returns to A, having taken approximately 224 days for the whole revolution. The line SA, SB, etc, from the Sun's centre to the planet's centre is always the axis of illumination, and the plane through the centre of Venus at right angles to these lines SA, SB etc. of the axis of the Solar illumination gives the position of the terminator on Venus' globe, IDL at D, IRL at R, IAL at A and similarly at the remaining parts of the orbit traversed by Venus.

As Venus travels on its 8-month orbit ABCD round the Sun, it traverses the twelve signs of the Ecliptic by its heliocentric movement; that is to say, as seen from the centre of the Sun, it moves through all the degrees of the Ecliptic measured by producing the lines SA, SB, SC and SD from the Sun to the farthest point of the Heavens to which the plane of the Ecliptic extends; and this is divided into degrees in exactly the same way as we have divided up the Ecliptic into the Signs of the Zodiac, as we travel in the circle EFGH described by the line SE which is the Earth's distance from the Sun; and this also applies to every other larger concentric circle, extending to the farthest regions of the Heavens.

That inner circle ABCD, which is the orbit of Venus, provides a visual illustration of the planet's illumination, seen from the Sun. Let us suppose that at a point of its orbit, R (as seen from the Sun, this point, on the line

SRΣ, is at the 20th degree of Leo on the Ecliptic), a plane perpendicular to the plane of the Ecliptic and passing through the Sun is set up, whose intersection with the Ecliptic, which is also the plane of our diagram, is the line ΣRSV. In that same plane perpendicular to the Ecliptic and passing through the Sun let us suppose the axis of Venus' rotation KZRM lies. In the other points of the orbit ABCD, that same axis of the planet's spin or rotation round itself will be inclined in the same way, parallel to KRM. For we have found that this inclination occurs around the 20th degree of Leo from observations to be recounted below. From these observations too we have deduced that the extreme points of the rotation of Venus round its own axis, that is to say the poles, do not lie on the plane of the Ecliptic at the points K and M. One of them, which we call North (pointing to the 20th degree of Aquarius, at V) is raised about 15° above the plane of the Ecliptic, and the South pole of Venus' rotation is depressed by the same amount below the plane of the Ecliptic at point M, which looks towards the 20th degree of Leo at point Σ.

The consideration of the illumination and appearance of Venus from the Sun in this diagram should have no intrinsic difficulty to those who, in serving their apprenticeship in astronomical studies, have become accustomed to the explanation with similar models of how not only the principal planets but also their satellites have always one half of their globes illuminated by the Sun.

Some explanation would seem to be necessary of the appearance of the part of Venus lit by the Sun, from the viewpoint of an observer not on the Sun but on the Earth. To such an observer that same planetary globe shows differing phases, now appearing like a small crescent Moon, now half and now gibbous, as it turns to us watching from the Earth, now a small part, now half and now more than half of its hemisphere lit by the Sun. But the explanation of this has become a standard feature in elementary textbooks on Astronomy, and indeed everybody knows the conclusions both in the system of Tycho and that of Copernicus. For in the system whereby the Earth moves round the Sun S, the line SA marks out Venus' orbit ABCD, and the line AE a third as long again marks out the Earth's orbit EFGH, divided by custom into 360 degrees through the twelve signs of the Ecliptic. Then if Venus is at point D for example, and a line drawn from the Sun's centre to Venus' centre SD, which is the axis of illumination, the orbit itself IDL divides the hemisphere lit by the Sun, IDLXI, from the dark half IDLYI. For at Venus' distance from the Sun, SD, which equals at least 4,000 diameters of the planet's globe IL, IDL can be considered as identical with the plane meeting the axis of illumination at right angles. Continuing therefore the plane IDL right to the Earth's orbit at Π, for example, the line DLΠ will be the axis of vision of the hemisphere YLXD which a spectator situated at Π will have, and because of which he will see Venus at dichotomy, since of the lit hemisphere IDLX he will only see half, DLX, and of the dark hemisphere LDIY he will only see half, DLY.

In the system of Tycho[11], however, the same diagram serves equally well to explain the phases of Venus in a similar way. But instead of the circle EFGH whose centre is the Sun, another circle of equal radius ΠS must be drawn from the point Π and similarly divided into 360 degrees, so that the Sun can complete its annual course through them, carrying with it as an epicycle the orbit of Venus ABCD, equally divided into 360 degrees. Then truly the same scheme of illumination and phases will also be demonstrated in this system too; but a much larger diagram than the preceding one must be used. For at the points F and H, to which the Sun's orbit reaches if the stable Earth has its centre at S, we ought also to extend the epicycle or orbit of Venus by a distance equal to DS. Therefore while only a third of the distance DS, namely HD, has to be added to the Copernican system, four spaces equal to HD have to be added for Tycho's system. Thus while the whole width of the Copernican diagram from F to H is divisible into eight of these parts, fourteen would be necessary for Tycho's system.

Here, therefore, for the sake of economy we prefer to use the smaller diagram of the Copernican system, on the understanding that the same phases apply equally and can be explained with equal ease in the diagram for Tycho's system if it is increased by the additions already indicated. I thought it necessary to point this out, in case anyone should deem that this view of the phases favoured one theory more than the other.

Above all an explanation is necessary of the arrangement of the markings, and how they describe their revolutions on the planet's surface, so that their paths, the axis of rotation and its inclination to the Ecliptic and its direction relative to the fixed stars can be deduced from observations. In pursuing this, however, we will follow the example of earlier astronomical treatises in using diagrams which can be applied to either theory; but in selecting the more compact one, we believe our exposition will be easier to follow, as everyone knows it also fits the larger Tychonian diagram.

Let us imagine, therefore, that Venus is at D while the Earth is at Π, to which point we suppose the plane IDL produced at a right angle to the axis of solar illumination SD. Venus should appear half, or at dichotomy, like the First Quarter Moon, to an observer at Π. If Venus' axis of rotation were perpendicular to the plane of the Ecliptic, spectators at Π could see all the markings on the planet during a single rotation. For the circle of Venus' equator (by this I mean the greatest circle, equidistant from both poles) would lie on the

---

[11] *A modification of Ptolemy's geocentric system, in which all the planets revolved round the Sun but the Sun itself along with the planets and the Moon revolved round the stationary Earth.* **S.B.**

plane of the Ecliptic, LXIYML. As the rotation progressed the quadrants LY, YI and IX would be carried in succession to the position occupied by LX, exposed both to the Sun and to our gaze at the same time. So a complete description of the globe could be achieved by us during the course of one revolution, in which the progress of the markings would be shown by straight lines parallel to the Ecliptic, because we would be looking at those circles from point Π on the same plane of the Ecliptic. On the other hand, if Venus' axis of rotation actually lay on the plane of the Ecliptic and was itself the line YDX prolonged to the Sun S, which we call the axis of illumination, then a single rotation of the planet would only show us half of the globe, namely the hemisphere IDLX, lit by the Sun. For the other hemisphere ILYI, although exposed to us in the same period of rotation, would not be lit by the Sun's rays and would arouse no sensation in our eyes. In such a disposition of the axis, the paths of the markings would appear as straight lines parallel to each other and at right angles to the Ecliptic. Likewise they will be straight lines parallel to each other but inclined to the Ecliptic, if the axis of rotation should be in the plane YDS but tilting above and below the Ecliptic. But if the axis of rotation were on the Ecliptic's plane, but pointing towards us through the line IDL prolonged to Π, the paths of the markings would appear as circles to us in that we would be situated on their axis of rotation.

In whatever other position the axis of rotation might lie apart from those mentioned above, with Venus at D and the observer on the plane IDL continued to Π, it is inevitable that the circles described by the markings as they revolve round the axis should appear as ellipses, according to the theory of conic sections. For the lines drawn from the eye of the spectator at Π to the circumference of any circle described by the markings form a cone. The intersection of the plane oblique to the axis with the circles made round that axis, cutting through the cone itself on all sides, produces an ellipse.

By changing the position of Venus from D to R, the spectator at Π does not see half of that hemisphere LKI lit by the Sun, but less than a quarter, depending on the size of the arc Lξ. But the circles described by the markings during their rotation, which will appear as straight lines for a spectator positioned at ψ, will show an elliptical form to our other observer remaining at Π, as indeed will the terminator itself. For this reason, the observer at Π sees Venus like a little crescent Moon, whilst the other one at ψ sees it at dichotomy.

Let us finally consider the case of a spectator positioned at Π, with Venus at R, where the axis of rotation MRK lies on the plane passing through the Sun's centre, S, and ask what difference in appearance the ellipses will have if the poles lie on the plane of the Ecliptic at points K and M, from the aspect produced if the North pole is raised 15° (for example) above the plane of the Ecliptic, and the South pole depressed below by an equal amount.

The theory of the analemma[12] will easily solve this problem, if first of all in this diagram we show the appearance which the parallel paths of the markings in both of these different positions of the axis would have for us, if we were at the pole of the Ecliptic looking down at right angles to their plane.

It is clear that in the first disposition of the poles KM lying on the plane of the Ecliptic, the circle of the equator IRL as well as the other parallels will be seen as parallel straight lines. But if the elevation of the North pole above the plane of the Ecliptic be placed between K and R, 15° above K, its appearance is thus deduced from the theory of the analemma. On both sides of point K let two equal arcs be marked of 15°, Ka and Kb, and connect a and b with a straight line. The half-diameter KR will cut it at Z. Point Z represents the North pole's elevation of 15° above the Ecliptic's plane. Let equal arcs of 15° be marked from I at e and from L at u. If the points e and u are connected by the line eTu, the point where this line cuts the diameter MK at T will be on the equator, 90° from the pole Z. A half-ellipse drawn through the three points I, T and L, with IL as its major diameter and TR as its minor semi-diameter, will show us, situated at the Pole of the Zodiac, the appearance of Venus' equator 90° from the pole Z, as it stretches across that quadrant. If there are any markings on the Equator, we will not get a view of them unless they are at the points I and L reached by the Sun's rays. For the whole arc ITL will lie in shade, not illuminated by the Sun. Again let an arc of 15° from I to n be marked and from L to X, and through the three points nRX a half-ellipse be drawn. This will give us the parallel described by markings 15° from Venus' equator towards the North pole Z. To us watching from the pole of the Zodiac these markings would seem, when first appearing on the hemisphere turned towards us, namely at point n, to be distant from the terminator by the arc nI. Then after six days on reaching R they would seem to touch the terminator, and again after another six days to recede from it for the distance LX when traversing the second quadrant TL.

But now let the observer's eye descend from the Pole of the Ecliptic, where up to now we have imagined it to be, to the actual plane of the Ecliptic, to watch the same markings revolving on the line nRX from the point ψ lying on the plane of the Ecliptic in the direction of the line IRL. When a marking, for example, 15° north of Venus' equator enters the hemisphere facing the observer, RZKXLMR, at its first appearance it touches the terminator R, then gradually rises above it until it has completed a quarter of its revo-

---

[12] *An orthogonal projection of the celestial sphere on the plane of the meridian, used in making sundials.* **P.F.**

lution RX. At X it has elongated from the terminator, and recedes even further when accomplishing the next quarter of its revolution.

Therefore two conclusions are to be deduced from this. The first is that, situated as we are on the plane of the Ecliptic, we can prove from the track of the markings when it forms a straight line or a curve parallel to the terminator; I repeat we can prove that the axis of revolution then lies in a plane passing through the Sun, and so find out the point in the Heavens to which Venus' axis of rotation points. The other is that we can also deduce from the size of the ellipses described by the markings approximately the degree of elevation of Venus' North pole above the plane of the Ecliptic.

Similar to this was the discovery of the Sun's axis of revolution by observing the tracks of sunspots, which rotate along with the solar disk in 28 days. In the case of sunspot observations, however, the experiment is much more exact and easy, since with a telescope of eight or ten palms the Sun's disk displays itself in a projection box as half a degree and three minutes in size so that even the tiniest spots and their daily progression can be recorded most accurately. But on the disk of Venus, which hardly subtends an angle of one minute when seen from the Earth even at its nearest to us, the spots which are revealed to us are faint, like the lunar ' maria' seen from Earth with the naked eye, and cannot make their own impression on a card held beyond the focus of the eyepiece when the telescope is turned towards the planet. The distance of the tip of each marking from the terminator has to be estimated, or a micrometer used (which hardly allows the distance to be measured accurately by its threads since the image moves past them so quickly), or the size of the markings has to be judged by comparison with the planet's actual diameter, which does not measure the exact number of degrees. Therefore it will be a fair assessment of our efforts if we claim for our observations a standard of precision that admits an error of four or five degrees in the measurement of angles and arcs of such tiny images, which pass rapidly across the field of view of telescopes of about 100 palms[13], the use of which is necessary for these observations to show the markings clearly.

After dealing generally with these theoretical considerations I must proceed to the diagrams of the tracks of the markings produced from the observations we made from February 9th to March 1st, which will be dealt with in Chapter 4. For now it suffices to know that we were unable to complete a description of the whole globe of Venus from these February and March observations alone, since from February 9th to March 1st perpetual shadow covered whatever lay on the planet's globe from the South pole through the arc of the great circle ML to the latitude of 57°. The proof of this I will give.

---

[13] *About 73 Imperial feet.* **P.F.**

On February 9th Venus, as seen from the Sun, was in the 17th degree of Cancer at D, and on March 1st it had progressed to R at approximately the 20th degree of Leo, as is shown from the tables of heliocentric movements of that planet. Around the 20th degree of Leo, as we will show in Chapter 4, the plane at right angles to the Ecliptic through the axis of Venus' rotation, namely ZRg, passes through the Sun's centre. Therefore on February 9th when the centre of Venus was at the 17th degree of Cancer, the axis of revolution, which always points the same way at different parts of the orbit, should be represented by the line MDZ parallel to MRZ. When the straight line DS crosses the parallels MDZ and MRZ it forms alternate equal angles ZDS and DSR. The angle DSR is 33° when it goes from 17 degrees of Cancer to 20 degrees of Leo. Therefore the angle ZDX, or the arc KX, will be 33°, and if to that arc is added a quarter of the same circle XDL, 90° (which is counted from the axis of solar illumination DXS to the great circle IDL, the terminator) the arc KXQL of the globe D will be 123°, the remaining arc LM of the semicircle will be 57°, which is always in the shade when Venus is traversing the arc DR of its orbit; and that is what we had to prove.

Since therefore a portion of Venus' globe, 57° from its South pole, in whatever direction it pointed during these months, could not be seen as it was not lit by the Sun, we had to wait until the planet had advanced by at least a quarter of its orbit from R, namely when Venus, as seen from the Sun, was at the 20th degree of Scorpio (which happened in that same year 1726 on April 27th). For there the axis of solar illumination fell at right angles to the axis of Venus' rotation, so that both poles were lit, making Equinox on Venus so to speak, and all the markings could be observed successively as they rotated. In the following days the illumination of Venus' South pole increases, and continues for four months until it reaches the opposite part of the Zodiac to the 20th degree of Scorpio, that is the 20th degree of Taurus, where again Venus enjoys Equinox and both poles are lit. Then from April 28th the South pole is again hidden from the Sun, and the North pole emerges.

Therefore in May and June we found we had to move on to new observations, to see the southern region of Venus then lit by the Sun and visible to us at the morning elongation, and describe the spots appearing on it, thus completing the Celidography of the whole globe. Thus by accomplishing this we completed in May and June what we had started in February and March, a description of the markings of the whole globe.

But in saying that we completed it, I would not wish astronomers to suspect that I was so confident about this first attempt that I would publish as a thoroughly completed work this preliminary investigation into the planet's globe and its markings. Rather I would willingly concede the possibility of corrections being made in the future after a second series of experiments repeated by us or undertaken by others.

The completely accurate description of which I speak needs a comprehensive survey which I could not perform from the observations of February 9th to March 10th, as nearly a third of the whole globe was hidden from our sight, with the South pole lying in shadow as far as 33° latitude, as I said. This South pole of Venus and area of 33° extending towards the equator was clearly shown and revealed to us in sunlight in May and June, and this allowed us to describe the whole planet's globe, even the southern markings; but it was not possible for us to investigate their smaller features and distinguish the faint edges of their curves and contours thoroughly and clearly enough and with sufficient accuracy to make further more detailed studies superfluous. Moreover, we hope and urge that more skilful observers should apply their efforts to a more accurate delineation of the 'maria' ("seas"), particularly those in the South. For since in May and June 1726 Venus was where the Ecliptic ascends less steeply and at the end of June was receding further from the Earth than it had been in February and March, it was more difficult for us to observe it for any length of time in the morning twilight, and because of the increased distance from Earth, the shapes and edges of the markings appeared less clearly defined, according to the law of the propagation of light, which diminishes according to the square of the distance. This is also experienced with Jupiter's bands. For in telescopes of 25 palms[14], we can see them clearly defined and count them when Jupiter is in opposition, no more than five times the distance of the Sun from the Earth. But it is much more difficult to see and distinguish them in the same telescopes after the planet's quadrature, when Jupiter is at a distance six times greater than the Sun's, and when it is approaching superior conjunction, seven times this distance from us.

So for these reasons I do not claim to give a completely accurate account of the whole Celidography of the planet Venus, particularly regarding its southern hemisphere. I offer the readers as much as we were able to achieve, leaving room for further studies, or rather even pleading that other more skilled observers should turn their attention to refining them further and correcting them by subsequent experiments.

VIII. We put off, however, to the last chapter of this little treatise the discussion of suitable times for attempting new experiments, as it is more fitting to speak about repeating observations after we have thoroughly dealt with, one by one, the measures we have taken so far to determine the planet's celidography.

---

[14] *About 18 Imperial feet.*

Meanwhile it suffices to say that I was confident enough about the parallelism of Venus' axis to be able to declare that the North pole would be favourably presented to us in June the next year, 1727 (as I write it is October 1727), then in March 1729, May 1732, and the South pole throughout June 1729 and October 1730. Again from mid-February to March of the year 1729 the northern hemisphere will present a similar aspect to us as at the first observations of the year 1726. Corresponding phases will also be seen in 1734 by those who turn their telescopes on it for the whole of February and the first part of March, when after an 8-year period Venus traverses, on almost the same day of the civil year, the same degrees of its orbit and the Zodiac, and forms the same triangle with the Sun that it did 8 years before.

That our hopes and expectations were fulfilled in July 1727, our observations carried out then and faithfully reproduced in Table III bear witness. For example on July 7th we used the usual 94-palm telescope made by Campani, set on the hills at Albano, and turned towards Venus at evening twilight, when after heavy rain the Mistral wind arose opportunely and cleared the air of all mists. Venus, like the gibbous Moon, appeared as shown in the diagram labelled July 7th in Table III. The North pole was at S, in fact on the plane passing almost through the centre of the disk and the cusps K and M. I saw a marking in the shape of a semi-circle there, whose tips n o and p r projected almost equally from the line of the plane KSM towards the bright part of the disk, X. Those tips n o and p r were a little thicker than the middle part of the marking at z.

On July 10th the telescope was turned on the planet again in the evening twilight half an hour after sunset, and I noticed that the tip n o had moved forward by almost an eighth of its circle beyond the plane KSM, and the other tip p r had moved towards x for an equal distance as n o had moved in the opposite direction during these three days. And yet the marking still preserved the same shape, that is a semi-circle like the letter C but curved in the opposite way, ⊃, as seen through the telescope, and the tips n o and p r were thicker than the middle of the semi-circle, z.

I brought the same 94-palm telescope to Rome, and on July 18th set it up in the Farnese Gardens on the Palatine Hill at the same twilight hour and, directing it at Venus, saw what I had expected, namely that the semi-circle marking n o z p r (shown in Table III, drawing for July 18th) was almost the opposite way round on Venus' disk from its position of July 7th. That is to say the tips n o and p r projected beyond the plane KM, not toward the bright part of the disk x as on July 7th, but toward the dark part and reversed, so that inverted in the telescope the marking looked like the Latin letter C. This was the only difference. The tips n o and p r were thicker than the middle z, as observed also on 7th and 10th July. So eleven days was the required time span to achieve this situation, since it is half the time of 24 days required for the planet to rotate around its own axis. So I decided to call this semi-circular marking near Venus' North pole Mare Boreum (North Sea) or the Sea of Marco Polo.

In subsequent observations, repeated after that date whenever time allowed, I discovered by studying a series of revolutions in order, that the edge of the semi-circular and circumpolar marking p r was on the same line of longitude or meridian (that is a great circle through the poles and equator) where, at about 38°, Mare Primum, or Mare Regii Joannis V (the First Sea, or the Sea of King John V) is to be found, the 38° line lying slightly beyond its middle portion. The other end of the semi-circle n o is at about 255° longitude, stretching a little beyond the western edge of the marking, to which I have given the name Mare Quintum or Mare Columbi (Fifth Sea, or the Sea of Columbus).

So these observations allowed the whole planet's globe and markings to be revealed, since it had been possible to determine the position and extent of each one seen from February 1726 to August 1727. Therefore the appearance will be the same eight years later, if the surface of Venus remains unchanged, as it is reasonable to suppose. But for other opportunities of observing Venus' phases and markings during this eight-year period, consult the last chapter of this small work, since the information given so far is sufficient to indicate how much our efforts have achieved already, and the additions we desire to the existing Celidography. Now, I will speak about this name 'Celidography' that I have given to this description.

IX. If anyone should enquire about the use of this word. I will reply that it is derived from the Greek expressions; Κηλις means 'mark' and γραφειν is the verb 'to write'. I decided to follow the example of those who have made maps of lunar markings and used Greek to call it Σεληνογρφια or Selenography. The Greek word Κηλις corresponds to the Latin 'macula', and Κηλιδοω is the verb derived from it like the Latin 'maculo', and Κηλιδωτος corresponds to 'maculatus' and 'Ακηλιδωτος to 'immaculatus'. In this sense, however, the Greek word Κηλις does not indicate a blemish, but rather an obscuration on the outside of a body contrasting with the other part that is bright and shining, when the strong reflection of light is impeded, which does not happen on other parts of the same surface free from the obscuration. It is like a mirror with rusty patches on it, or broken by cracks, like marks on its polished surface, preventing an equal ability to reflect light. Even in Holy Writ, e.g. the Book of Wisdom, Chapter 7, Verse 26, the Greek reads 'εσοπτρον 'ακηλιδωτον, translated into the Latin Vulgate as "a mirror without blemish". These interruptions of the brighter light by more obscure or paler areas have long been noticed on the Moon and other planets. So now they are also on Venus, where we shall call the markings Κηλιδες. The word is Κηλις in Attic Greek, but in other Greek dialects σπιλος, as is explained in the Lexicon of Crinitus: "σπιλος means stain, blemish, pollution, birthmark, a mark on the surface; in Attic Greek it is not σπιλος but Κηλις, according to Phrynicus." Plutarch[15] calls the lunar markings σπιλος in his work "Concerning the Face on the Moon's Orb", p.921 in the Greco-Latin edition. Latin and Italian translators render this as "large black spots", called Κηλιδες by the Attic Greeks, and which Plutarch in the aforementioned work says many philosophers think are seas. Others think they are images of seas on the Earth, reflected on the Moon as in a mirror. Since therefore the Latin version of the word Κηλις is used to describe the lunar markings, it can also quite aptly be used for marks observed on Venus, which when seen through 100-palm telescopes arouse similar sensations to the lunar ones seen with the naked eye, that is to say without a telescope.

X. Since my account of Celidography has highlighted this similarity of appearance of the markings of both Venus and the Moon, the former through 100-palm telescopes, the latter with the naked eye, it will be relevant to my subject to examine the reason for this great similarity of aspect.

Therefore I purposely made observations on March 7th at Rome, May 25th at Albano and June 22nd at Rome, when Venus was near the crescent or half-moon, in a telescope of 94 palms, about the required size, fitted with eyepieces of 7 $1/2$ or 10 unciae. I saw Venus as a crescent, near to the Earth, and of the same apparent size that the Moon was with the naked eye. It is very easy to show how this was possible. Venus as seen by us that year around March 1st was at about $2/5$ the distance of the Sun from the Earth. This can be shown by Trigonometry from the triangle formed from lines joining these three bodies, the Sun, Venus and Earth. Of this triangle every angle is already known, but on the diagram in Table IV, which I drew in accordance with the measurements, it will suffice to use dividers for the sake of ease. Now this distance between the Earth and the Sun I have found to be about 11,200 times the Earth's radius, as I will show in Chapter 7, where the subject will be the observed parallax of Venus. Therefore Venus was about 4,500 times the Earth's radius distant from us on March 1st. The Moon's average distance from Earth, half-way between Apogee and Perigee, is 58 times the Earth's radius according to very accurate measurements of modern astronomers. Nor do the ancient investigators disagree greatly, since in Plutarch the average distance of the Moon from the Earth is given as 56 times the

---

15Greek biographer and philosopher of the first century A.D. **S.B.**

Earth's radius. Therefore on March 1st the distance of Venus from the Earth was almost 80 times greater than the Moon's average distance, and almost 90 times greater than the minimum distance of the Moon, which is agreed to be 54 times the Earth's radius. If therefore the telescope gives a magnification of 90 or 100 times (as a slightly larger magnification would act as compensation for the diminution of light by distance), the image of a mark on Venus in that telescope will have the same size and intensity as one on the Moon of equal size seen without a telescope. The telescope we used, however, gave a magnification of 112 times, with an object-glass of 94 palms focal length or 1128 unciae, and an eyepiece of 10 unciae. So this is the reason why Venus' markings arouse a similar sensation in the eye seen at that distance with a 90 or 100 palm-telescope, and eyepiece of 10 unciae, as do the lunar 'maria' at the Moon's average distance from the Earth, seen with the naked eye by people who enjoy normal eyesight. Moreover, even the appearance of Venus at dichotomy as at the beginning of February and June, although it was a little further from us than on March 1st, about 6,000 times the Earth's radius, should have been similar to the Moon seen with the naked eye, in the aforementioned telescope. For the average distance of the Moon from the Earth of 58 Earth radii multiplied by 112 gives a distance of 6496 Earth radii for Venus at that time. Through a telescope of 94 palms or 1120 unciae fitted with an eyepiece of 10 unciae the image was magnified 112 times, increased to the size that Venus would appear to the naked eye if it were transported from its present distance to that of the Moon.

When I said just now "by people with normal eyesight" in connection with observing lunar markings with the naked eye, I was referring to an old problem raised by Plutarch in his renowned work "Concerning the Face on the Moon's disk". Since this has considerable bearing on a trustworthy account of the experiment we made concerning Venus' markings being put forward, and also on the guidance of attempts by others in the future wishing to observe and compare their findings on Venus' markings with our attempts at Celidography, it is appropriate to quote Plutarch here.

At the beginning of the work I mentioned, a certain philosopher is introduced by Plutarch and remarks that these larger lunar markings which are called 'maria' and popularly are thought to display a certain shape, or represent a mouth and eyes like a human face on the lunar disk, are not obvious to people of poor eyesight, to whom the Moon's disk does not appear to be varied in colour like this. He asks moreover the reason for such defective vision: "Why do people of poor and weak eyesight see no variations of shape on the Moon, but its globe shines with even brightness for them, while people of keen eyesight discern subtle variations of shape and note these differences with greater clarity?". During the course of the same work, however, he gives the sizes of the markings distinctly seen by people of keen eyesight as half a digit, having divided the Moon's disk into 12 digits. But this gives remarkable proof that these larger lunar markings, called 'maria' by the philosophers Plutarch mentions and also by contemporary astronomers, could not be seen clearly by those described by Plutarch as having very good eyesight. For telescopes show us that several of these markings or 'maria' are not only larger than half a digit of the lunar diameter, but are even a third of the whole diameter. So if the markings of Venus appear almost the same size as the lunar ones do to the naked eye, when seen through a telescope magnifying a hundred times, it is not to be expected that they will appear clearly to everyone's eyes, since in measuring lunar markings of this kind, eyes which were thought to be so acute and keen were guilty of such a margin of error. It is not inappropriate to give this warning in advance, so that when looking at Venus' markings measures should be taken to select observers of sufficiently keen eyesight to be able to make out the larger lunar markings called 'maria' without the aid of a telescope.

Here I will also suggest a reason why the visibility of the markings on Venus was much more obvious in February and March than around mid-June, and why their colour became much paler and weaker at the end of June. For from mid-June to July 1st Venus' distance from Earth was almost double what it was in mid-February. Just as the lunar markings would hardly arouse any sensations in the eyes of observers not armed with a telescope if the Moon were twice as far from us as it is now, since at its average distance markings cannot be seen by those with poorer eyesight, so also Venus' markings, 90 times further away from us than the Moon, seem similar to the lunar ones in a telescope magnifying 100 times, but when twice as far away, that is 180 times further than the Moon, they hardly show even a weak and pale image, just like the lunar ones if removed to twice their distance.

After issuing these warnings of precautions to be taken in the selection of times, equipment and observers to undertake these experiments, we will proceed to demonstrate the Celidography or mapping of the markings detected on Venus. Half of these we saw at the evening apparition in February and March, when Venus bore the name Hesperus, and they revolved round the axis, which about the 1st of March lies on the plane drawn through the Sun and cutting the plane of the terminator at right angles, and from the North pole to about 20° beyond the Equator they were exposed to the light. The rest we saw in May and June, when it was called Phosphorus at its morning apparition, and its axis of revolution, which keeps the same direction, showed the South pole to us and to the Sun, also a considerable part of the northern Celidography already inspected. The complete southern hemisphere of the planet was revealed to us during the course of the rotation in 24 days, although it was furthest from us when it started to rise before dawn at its morning apparition. Thus the boundaries of the markings were less clearly defined as the intensity of the light fell off. Therefore the shapes of the markings beyond 20° S up to the South pole we wish to be determined more accurately from a closer distance some time by future observations either by ourselves or others, in order to give a more faithful and exact representation.

CHAPTER III

CONCERNING THE CONSTRUCTION AND USE OF A GLOBE, PLANISPHERE AND DEVICE TO SHOW MORE CLEARLY THE CELIDOGRAPHY AND OTHER OBSERVATIONS OF VENUS.

Summary of the Chapter

I. The construction of a globe, planisphere and armillary device to be based on the observations of Venus' poles, axis of rotation and equator.

II. The equator was seen clearly and almost coincided with or lay along the terminator on March 1st 1726.

III. This is shown from the diagram of observations.

IV. The point on the Ecliptic must also be sought which is crossed by the plane produced through Venus' axis of rotation and the Sun that is at right angles to the plane of Venus' eight-month orbit round the Sun.

V. On the diagram of the planisphere showing Venus' orbit round the Sun, with the eye of the spectator placed at the pole of the Ecliptic, this is explained more clearly.

VI. Observations made in 1726 in the month of February indicate that the plane produced through Venus' axis of rotation and the Sun meet the Ecliptic at about the 20° points in Leo and Aquarius.

VII. After determining the point of the Ecliptic which cuts that plane which represents the solstitial Colure of Venus, and showing the amount of 15° by which the axis of rotation is raised at the North pole 15° above the plane of the Ecliptic and depressed 15° below at the South pole, the construction of an armillary device is proposed to show all Venus' phases.

VIII. That armillary device can be adapted to both systems, that of Tycho and that of Copernicus.

IX. The same effect achieved through a planisphere, the construction of which for either system and use is demonstrated.

X. The use of another little machine to be fitted to the planisphere is explained, so as to represent accurately the different phases and markings to be seen on Venus.

I. To give a clearer idea and fuller understanding of the disposition of Venus' axis of rotation and its markings with respect to the plane of the terminator and also by comparing our observations of the markings with others to be undertaken in the future, to enable a judgement to be made on whether any changes of the planet's topography have occurred, we have had a solid globe made to represent the planet Venus. After marking the poles and axis of rotation on that globe (as geographers and astronomers are accustomed to do on the terrestrial or celestial globe and the armillary sphere) we have indicated on its surface the markings observed by telescope, retaining as far as possible their relative sizes and distances from each other and from the poles and greatest circle of rotation between them which we call the Equator.

II. From the orderly progress of the markings which we observed daily from February 9th, we discovered that on the days just before and just after March 1st the circle called Venus' equator did not exactly coincide with the planet's terminator, but the plane of one of the circles was slightly inclined to the plane of the other. For the parallels of rotation described by the tips of the markings were not completely equidistant from the terminator but nearer to it when starting to appear on the hemisphere turned towards us and gradually becoming further away, as if they had not arisen on the dark side, the more they moved daily towards the other cusp of the hemisphere. It was noted that they approached it 12 days after their first appearance, and 6 days after occupying the middle of the hemisphere turned towards us, or the centre of the disk.

III. Look again at the diagram of observations that we referred to previously, Table II. Compare on it the appearance and progress of the three-part marking observed on February 14th, 16th and 18th. The tip of the middle section 3, which extended further than the two smaller ones on each side, 2 and 4, occupied on February 16th the centre of the disk and the hemisphere turned towards us. Two days before it was nearer the cusp S from which it had begun to appear, and two days later it had moved forward by the same amount towards the other cusp, R. The same thing happened with the tips of the other two on either side, keeping the same motion and describing similar parallels. If parallels drawn from the tips were equidistant from the terminator SCR, both in the middle of the disk and beyond the middle, they would be seen to move equally. But on the 16th February the tips of the markings were more elevated above the terminator than they were on the 14th. Then on the 18th they were still more elevated, and after them came the lower tip 5 of the other following marking E5, which on the subsequent days was observed to be rising above the terminator. It was necessary therefore to conclude that the parallels described by the markings, and the equator of

Venus, were inclined a little to the terminator, and the axis of rotation did not form the same angle with both cusps of the planet at crescent phase, but slightly under a right angle with the cusp S where the markings started to appear, and slightly over a right angle with the other cusp R, to which the markings gradually moved. But since the axis of the terminator is the same distance from each cusp, and goes through the centre of the disk exposed to us and through the Sun, and lies in the plane of the orbit of Venus round the Sun, which is the Ecliptic, it follows that the axis of rotation rises by an equal number of degrees above the Ecliptic as the axis of rotation in the hemisphere of Venus exposed to us exceeds the middle point measured equally from each cusp. The distance from the terminator of markings rotating is greater as they disappear than when they appear, as we have observed, sometimes using a micrometer and sometimes the judgement of the eye, comparing with the bright area of the crescent Venus the parts which the markings make hazy or more dusky. This indicated an inclination estimated at about 15 or 20 degrees of either axis, that of rotation or that of illumination, to the other. Because of this, we concluded that on the machine intended to show the phases of the planet, one of the poles of rotation (which we shall call the North) should be raised 15 or 20 degrees above the plane of Venus' orbit round the Sun, and consequently the other pole (which we shall name the South) must be lowered the same amount below the plane of Venus' orbit, or the Ecliptic.

IV. The other question to be investigated was to which part of the Zodiac the plane through the axis of rotation and the Sun pointed, so that we could find out from that whether the axis of rotation kept constantly in the same direction as the planet traversed its 8-month orbit round the Sun and at the same time turned on its axis in 24 days. How I went about this will be explained on the diagram of the planisphere that I have now produced.

Let us now examine the diagram of the planisphere that I have drawn explaining the orbit of Venus round the Sun, and the changing aspects of illumination in the various parts of the orbit as shown in Table IV.

V. Let the centre of Venus' orbit be the Sun, S, round which Venus revolves in 224 days from A through B, C, and D, returning to A, traversing each quadrant AB, BC, CD and DA in 56 days. A line from the centre of the Sun, S, to the centre of Venus' globe is the axis of the circle of illumination. Let the centre of Venus be at one point of its orbit, for example R. The line SR from the Sun, S, to the centre of Venus, R, and prolonged to the far side of the planet, M, is the axis of the terminator IRL dividing the illuminated hemisphere IRLK from the opposite, dark hemisphere IRLM, and the two extreme points of this axis, K and M, are the two poles of that same terminator IRL, and lie on the plane of the Ecliptic, or Venus' orbit round the Sun.

We have seen above that Venus' poles of rotation or spin round its own axis do not lie on the same plane of that orbit, or the Ecliptic, K and M, but the North pole at Z is raised 15 or 20 degrees above it, and the South pole, g, is depressed below it. It is possible however for both poles of rotation to lie on a plane perpendicular to the Ecliptic, and through the poles of illumination, K and M, even though the poles of rotation be raised and depressed above the plane of the Ecliptic and the poles of illumination, K and M. For let us imagine a circle IMLK drawn, which represents a section through the planet's centre at the Ecliptic. Above the plane of this section at right angles let another plane be drawn through the Sun and the centre of Venus, the intersection of which with the plane of the circle IMLK lying on the Ecliptic will be the line MRK. To a person viewing the diagram from the plane IMLK the hemisphere of Venus above that plane will reach its northern apex at R, with the result that the point R, which formerly indicated the centre of the globe, will now indicate its highest point, and the line RK, which earlier represented the radius and half the axis of the circle of illumination, will now also represent a quarter of the circle curving from the highest point of that hemisphere, R, down to the plane of the Ecliptic at K. For if the eye looks at the diagram not from the pole of the Zodiac and Ecliptic but from such a position that the radius of the orbit SR appears as a single point, according to the theory of the analemma, from that perspective the circle through the poles of the Zodiac and perpendicular to the plane of the Ecliptic ought to be represented by a straight line, and the quarter circle RK, which is equal to IK and KL, should be shown by the line RK. On the line RK we mark the degrees of that quarter circle in the accustomed manner of the analemma. Supposing for example we have to mark the 15 or 20 degrees of the polar elevation above the point K, lying on the plane of the Ecliptic. From K in the direction of I an arc Ka of 15 degrees is marked, and an equal arc Kb in the direction of L. The points a and b are connected by a straight line aZb which cuts the line RK at Z. Z is taken as the elevation of the pole of rotation above the pole of illumination K, and 15 degrees above the plane of the Ecliptic in which K lies. Therefore in this position is the plane perpendicular to the orbit of Venus or the Ecliptic MRK, like the Solstitial Colure on the armillary sphere, on which we found the poles of Venus' spin or rotation around its own axis, represented by the points Z and g, and the poles of illumination, represented by the points K and M.

Now let us imagine that Venus has travelled to another point of its orbit, say D, where the line SD from the Sun to the centre of the planet marks the axis of illumination. At this point the plane through the axis of rotation perpendicular to the Ecliptic does not coincide with the plane SD, because (as we will show at the appropriate place) Venus' axis of rotation, like a magnetised bar carried around in a room, always maintains a direction that is parallel in the various parts of its orbit. Therefore let a line DK be drawn from the centre of Venus at D, parallel to the line KR, to show the parallelism of the axis. The plane MDK is the plane of the circle raised at right angles above the Ecliptic of Venus in which the point Z (as in the previous projection MRK) denotes the North pole of Venus' rotation, elevated as before 15 or 20 degrees above the plane of Venus' orbit or Ecliptic. When the line SXD from the Sun to the centre of Venus meets the parallel lines MDK and MRKS (common sections of the orbit of Venus or the Ecliptic with the plane at right angles to it through the axis of rotation) it will form alternate equal angles KDS and DSR. The angle DSR is known from the intervening movement of Venus round the Sun. Therefore the angle equal to it, KDS, will also be known, once it is established at what point of Venus' orbit the plane of the axis of rotation coincides with the axis of illumination.

VI. Therefore our examination will proceed to investigate the place at which these two planes coincide. When this is found, the full explanation for the illumination and the progress of the markings can be accurately given. Furthermore, this can all be deduced from the observations.

Let us therefore look at Venus from a different perspective in Figure 3 which I have added to Table IV. Previously we showed the plane of the Ecliptic, or Venus' orbit, and the eye of the spectator was placed at the pole of that Ecliptic. Now we put the spectator's eye on the plane of that same orbit, or Ecliptic, in such a way that the plane itself appears as a straight line MKS on this diagram, as happens according to the rules of perspective when the horizon appears as a line to the observer. On that line even the Sun S will be situated. From the Sun S to the centre of Venus R the line produced to M is the axis of illumination, and the plane perpendicular to this axis through the planet's centre R, namely BRA, is the plane of the terminator. The spectator's eye can be placed wherever one wishes on this plane perpendicular to the axis of illumination. If anyone placed at the centre of Venus were to see such a spectator, and the Sun situated on the Axis, a right angle would be formed between these two lines of sight. Indeed this is the angle formed when to us, situated on the Earth, Venus appears stationary. For a line drawn from the Earth to the orbit of Venus touches it at the point occupied by Venus from which another line drawn to the centre of that orbit (i.e. S the Sun) will form a right angle with the first line, namely the line of sight along which we watchers from the Earth see Venus at that time. Therefore from that position we see half of the hemisphere of Venus lit by the Sun, or a quarter of the whole globe, indeed the portion BRAKB; but the other quadrant BRAMB, that is to say half of the hemisphere not lit by the Sun, we would have been able to see also if it had received any light, since it is turned towards us.

But it makes no impression on our sight, since it is deprived of rays of light. Let us suppose that Venus is stationary at this position, and that its axis of rotation lies on the plane BMARKS, in which is the axis of illumination, KM. We have said that above this axis of illumination the axis of rotation is raised about 15 degrees. So let 15 degrees be counted from the point K situated on the plane of Venus' ecliptic towards the point B on that same globe perpendicular to the Ecliptic, and call this point Z. A similar amount should be measured from M towards the South pole A, and marked as g. The line ZRg will be the axis of rotation, around which markings and individual points on the globe will describe their parallels, and the Equator will display the greatest circle on the globe. This will appear to us at our situation as a straight line CRD, but other points will describe parallels somewhat curved, the more so the nearer they are to the pole, and they will appear to curve towards us when in the middle of the disk turned in our direction, according to the rules of perspective, if we observe Venus in this plane from close proximity. However, at the great distance of Venus from the Earth, this difference is hardly noticeable, and they can be considered as straight lines.

Suppose that Venus is at the same part of its orbit, and the Sun and the observer on Earth are in the same positions, but that only the disposition of Venus' axis of rotation is altered. Whilst in our first model that axis lies on the plane BMARKS at right angles to our line of sight, now in this second model let it be moved through a quarter of a circle and placed on the line of our sight in the plane represented on this diagram by the line BRA. If both poles of rotation are neither raised nor depressed above or below the plane of the Ecliptic, the point R as seen by us on Earth in that position will represent the whole axis of rotation and both its extremes, namely the poles. So the parallels described by single points on the globe of Venus through the spinning and rotation of the planet round its axis will appear as concentric circles round that central point R. The greatest of these circles, or the Equator, will be the circle which represents the planet itself, BMAKB, and markings by their own rotationary movement will display these concentric circles to us. But if the poles of rotation were raised 15 degrees and shown on the line BA by the points h and q, the Equator of Venus will appear as the ellipse ψMϕKψ, and the markings situated on the Equator of Venus and on the hemisphere lit by the Sun BRAKB, will display half that ellipse ψKϕ in 12 days. Markings situated beyond the Equator will describe smaller ellipses depending on their distance from it. All this can be proved from Theodosius' geometry of the sphere.

Finally a third situation of the axis of rotation should be considered, in which the axis of rotation is not in the plane through the line of sight, as in the second model, nor in the plane perpendicular to that, as in the first model, but in an intermediate place. It is obvious that ellipses will be described by the markings everywhere, and the diameter of each one can be shown and its extent from the notes on the geometry of the sphere referred to above.

So now that I have put forward this theory of the ellipses, circles or straight lines described by the rotating markings depending on their position on the globe of Venus and the position of the spectator observing their course, everyone can follow the reasoning by which we have tried, from our observations of the progress of the markings seen by us on the disk of Venus, to determine the place in its orbit where the axis of rotation coincides with the axis of illumination and passes through the Sun.

As far as I could ascertain from my observations, it appeared to me that this happened within a period of 10 days from February 23 to March 5. Markings situated on Venus' Equator displayed their circles as straight lines to us; that is to say, to us watching the planet from Earth, when it was not at dichotomy but at crescent phase. Taking account of this phase, I judge that it was around the 20th degree of Leo, as seen from the Sun, that Venus in its orbit pointed its axis of rotation in the same plane as the axis of solar illumination. Therefore the poles and the colure (so to speak) of the solstices of Venus sustaining the axes of illumination and rotation, are to be placed at the 20th degree of Leo and at the point opposite, the 20th degree of Aquarius.

That degree of the Zodiac at which the vertical plane of the axis of rotation through the pole of the Ecliptic coincides with Venus' axis of illumination could be defined more accurately if the movements of the markings on Venus' globe could be plotted with the same precision as those spots on the Sun's disk. There is, however, a vast range of difficulty between these two operations. For we see the Sun's disk in a projection box shining most clearly by its own rays and extending over half a degree of the great circle, and even the smallest spots are accurately defined in telescopes shorter than five or six palms, and we can record their daily progress on the same chart at whatever hour we wish, and measure them with compasses and ruler.

However, the disk of Venus at its nearest barely fills one minute of the great circle, and markings are only visible in telescopes of almost a hundred palms, nor can their images be projected onto a card. A disk of one minute passes quickly across the micrometer wires, even if magnified in telescopes of such length, so that a better attempt at recording markings can be made by estimating their positions relative to the cusps and limb of the crescent disk which is like a small moon, than by an accurate tracking of the courses of the markings across the wires of the micrometer. We ought, therefore, to define the place approximately, maintaining that it is somewhere around the 20th degree of Leo and Aquarius that the planes of rotation and illumination coincide at a right angle to the Ecliptic as Venus reaches that part of its orbit. We do not state that this happens exactly at the 20th degree of Leo and Aquarius, but about there, allowing for a margin of error of 10 degrees either way, namely somewhere between the 10th and 30th degrees of Leo and Aquarius. If sometime a more accurate method presents itself of tracking the movements of the markings on the disk of Venus, we will define the degree of intersection more confidently. At the moment our observations do not allow that point to be determined with greater precision. We will assume therefore that both these axes do in fact coincide in the same plane at exactly the 20th degrees of Leo and Aquarius, but are ready to make corrections if more accurate experiments demand it. We must move on to consider the construction of our machine.

VII. In addition to the inclination of Venus' axis of rotation with the plane of the Ecliptic at an angle of about 15°, we have discovered that the axis keeps parallel throughout the planet's orbit round the Sun, just like a compass needle is observed to do when carried round a room. When we include this phenomenon with the others already mentioned, the complete explanation of the phases of Venus and the markings appearing on its surface will be forthcoming. Therefore we will provide a demonstration in visual terms using two methods, a two-dimensional one using the planisphere, and a three-dimensional one using circles and armillary devices.

It will be appropriate to start our demonstration with the three-dimensional machine using circles and armillary devices. We are indebted to the ingenious craftsman Peregrino Mazza of Bononia[1] for part of this device, which shows the parallelism of the axis in a very clear and simple manner, the difficulty of imitating which has troubled quite a few craftsmen .

So let us set up at B (Table V, Figure 1) a rigid slender pillar AB to represent half the axis of the Zodiac, and place the Sun S in the centre, whose body and light are represented by a round lamp A placed in the centre. Since the Sun is the centre of Venus' orbit (really an ellipse but very nearly a circle) in both the Tychonian and Copernican systems, if the quadrant VE of a circle whose radius is AE or AV is attached to the slender pillar AEB at E and able to revolve around it, and at the end of this quadrant a small globe V is placed, the globe situated on the revolving circle VHE will describe the circle VPMOV around AE. This will represent to us Venus' orbit round the Sun, completing a full revolution through all the degrees of the Ecliptic in 224 days. But because it is necessary to show as well as Venus' orbit round the Sun, the planet's rotation or spinning round its own axis, and this axis inclined 15° to the plane of the Ecliptic and pointing in the same direction

---

[1] *Modern name Bologna.* **P.F.**

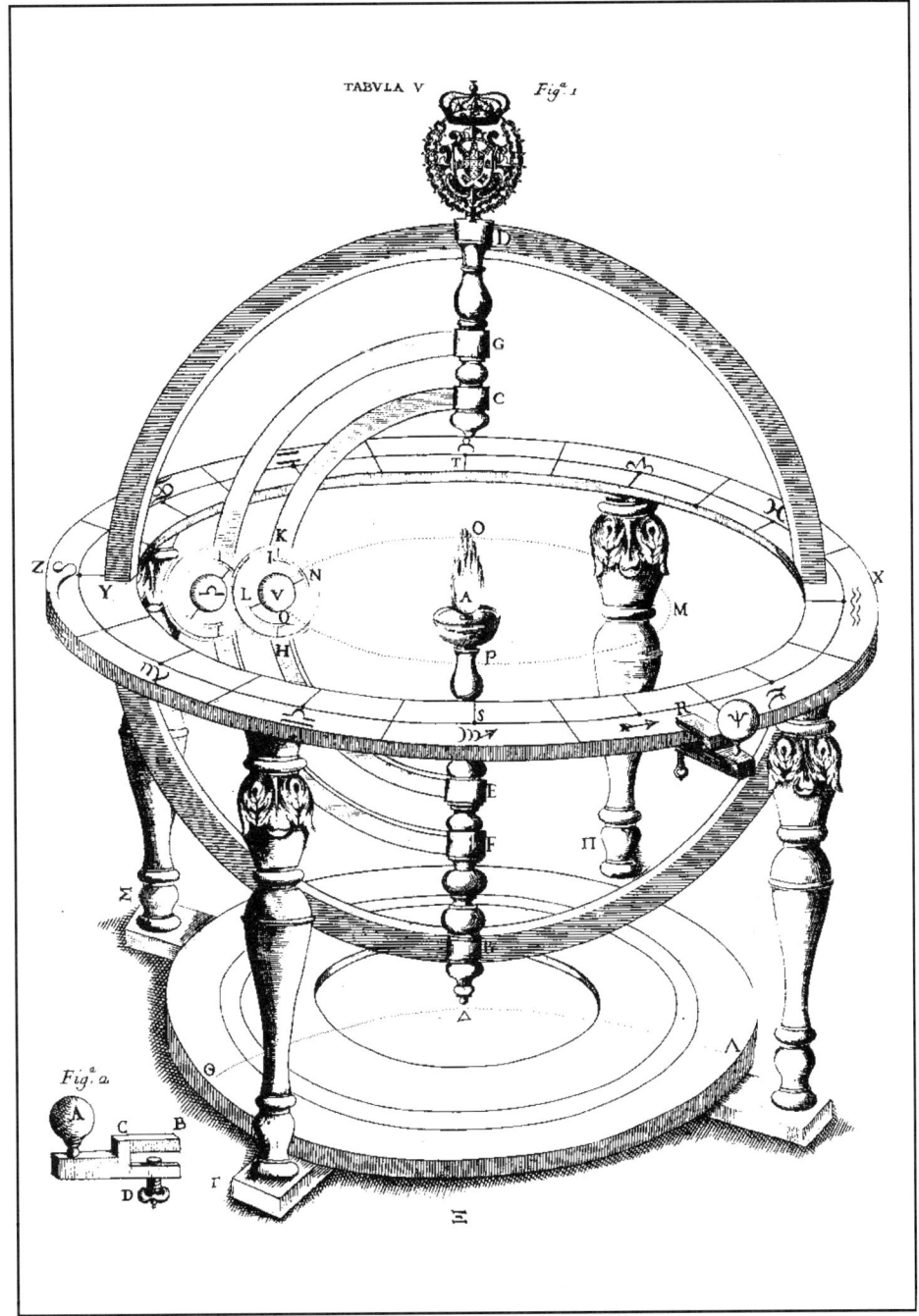

TABLE V

throughout the orbit, a semi-circle made of copper should be placed in the quadrant EHV continuing to C, and from C the central pillar should be continued to D in the direction of the first half-axis AB, showing the whole axis of the Ecliptic from B to D, with a gap however between A and B so that the lamp can be lit in the centre to show the Sun's rays reaching the planet in its orbit. Also the semi-circle CVE should have a bend in it like half an epicycle from K to H, so that there is enough room to include on the globe V representing the planet Venus a demonstration of its phases, rotation, inclination and parallelism of the axis. From the globe V two thin spikes VN and VL should protrude to show the axis of revolution or spin, inclined 15° to the plane of the Ecliptic XAZ. Some perforated, curved strips NI and LQ should be attached, segments of a circle from the centre V with radius VN or VL, forming equal arcs N-I and Q-L of that circle, each measuring 75°. At the points I and Q, curved plates hold pointed axles IK and QH, whose ends fit in to openings at K and H attached to the semi-epicycle KLH. This ensures that the axis of revolution or spin of the planet LN has its correct inclination of 15° with the Ecliptic and allows the globe to revolve around its own centre through the poles N and L. The axis LN constantly keeps the parallelism of its position through a whole revolution of the deferent circle CVE around the axis of the Zodiac DCEB by a device within the globe itself ingeniously contrived by the craftsman which will be explained at the appropriate place. This has the effect that, with the Sun always at A, the centre of Venus' orbit, when Venus is at the point of its orbit V, the North Pole of its rotation, N, is lit by the Sun, and the South Pole, L, remains dark. When Venus is carried round to P or O, both poles N and L are touched by the Sun's rays, and when it reaches M, the South Pole, L, is lit by the Sun and the North Pole, N, is devoid of light. By this distribution of rays over Venus' globe, which completes its orbit VPMOV in 224 days, we can understand how the two poles are lit alternately. Also, the presentation of the markings following each other is displayed, since the globe itself turns round LN, the axis of rotation.

It also remains for the direction of the parallelism of the axis LN to be related to a certain point in the Zodiac, discovered to be in the 20th degree of Leo and Aquarius, by the machine. This need is fulfilled by the half-armillary DYB, or indeed the whole armillary DYBD, which we must understand to be rigid and encircling point A, so that it can provide a means of displaying the Zodiac. Thus it will intersect with the 20th degree of Leo and Aquarius, the axis of the parallelism, or indeed the plane in which the axes of illumination and rotation meet.

The wider armillary or circle XSZTX should be prepared and supported on four firm legs, as is customary for the horizon to be shown on artificial globes of the Earth and the Heavens. To show the plane of the ecliptic, it should be set up to represent the line XAZ through the Sun A and at right angles to the axis of the Ecliptic DAB. It is obvious that all lines from the Sun A to any points on the orbit of Venus VPMOV and produced to the Zodiac XSZTX will show the intervening motion of the planet round the Sun and the degree of the Zodiac in which Venus is seen from the Sun. One of these points will be V, in which the plane prolonged through the axis of rotation LN cuts the Sun's centre at A. The same plane produced to the Zodiac cuts the Ecliptic at X. Therefore the point X should be fixed at the 20th degree of Aquarius, at which we see the South Pole of Venus' rotation L illuminated, while the axis of illumination and the axis of rotation coincide in the same plane AMXD. The opposite part of the Zodiac, Z, should be fixed at the 20th degree of Leo, where Venus shows its hemisphere containing the North Pole of rotation lit by the Sun. If the markings are shown in their proper places on Venus' globe (as we will explain below) the spin of this globe round the axis of rotation LVN will reveal the parallels of each one gradually curving from the dark hemisphere to the hemisphere lit by the Sun; the very same spots, and moreover at the same places, as are really visible in the sky.

VIII. It finally remains for us to decide the situation of the spectator on Earth to observe these phases.

With this armillary machine it is possible for this to be shown in a practical way which serves equally well for Tycho's system with a stable Earth, or for the Copernican system, where it moves. I stress this to prevent anyone thinking that these observations fit in with or favour one system rather than the other. The construction of the machine will just have to be varied slightly, depending on which of the two systems you decide to demonstrate. If the well-known Tychonian model is preferred, the centre of the machine will have to be put not at B but at Ξ, and the axis DAB, which holds the Sun, must not be fixed firmly in the plane BA but placed on the circumference of a circle ΘΔΛ, turning round its radius BΞ, the measurement of which we shall give shortly. Above the fixed centre of the machine which in this Tychonian system will be at Ξ, a pillar of the same height should be set as the one AB previously mentioned, and a globe at the top of it representing the stable Earth. In order that the plane through the centres of that globe and the Sun A might represent the plane of the Ecliptic, the machine fitted to the Tychonian system will have to extend much further and show the full circle of the Sun's orbit measured from the centre of the Earth to a distance equal to the line ΞΔ. The length of the line ΞΔ or the radius of the Sun's orbit on this machine will be dependent on the distance chosen for the radius of Venus' orbit round the Sun, AV. It is known from the observations of all astronomers that the radius of Venus' orbit round the Sun is in the ratio 3:4

with the Sun's orbit round the Earth. Therefore if MA or AV is divided into three parts, four of these parts will give the radius ΞΔ, and the circle described by this radius, ΘΔΛ, will carry the pillar BA round the centre Ξ, along with the half-armillaries fixed to it, BYD and EHVKC; while the point B is carried round the circle ΛΔΘ, the Sun's centre A will describe a circle parallel to this from X through A to Y; and in the plane of the Ecliptic, extended further beyond SAT from the centre of the globe above Ξ, at the height of the pillar BA, keeping upright, and representing the fixed Earth, a line through the Sun A will show the degree of the Ecliptic in which the Sun is seen. However, the circle EVC, carrying the globe of Venus V, while turning round the axis FA, will have to be positioned at a given date where a straight line from the centre of the Sun A through the centre of Venus V and prolonged to the Ecliptic will meet the correct degree of the Ecliptic for that day of Venus' orbit round the Sun, always taking into consideration the point we made previously, that the strips IN and LQ must be placed in the same plane as the semi-armillary DYB when the latter is on the straight line AVY prolonged to the 20th degree of Aquarius . With these provisions, all the movements of the planet Venus and its markings which are seen by the observer on Earth can be shown to spectators in accordance with the Tychonian system, through the three-dimensional medium of the armillary machine.

If, however, it is wished to display these same movements and phases of Venus and its markings according to the other system whereby the Earth moves, a much more compact machine will serve the purpose. For it will not be necessary to have as wide a circle for the Ecliptic as in the Tychonian machine. But its construction according to the measurements which I will give for the separate parts will serve very well to show all the planet's phases and markings very clearly.

The copper pillar AB should be a foot long, namely 12 unciae[2], as in a Roman foot. This serves as a radius of the metal semi-armillary or semicircle BYD, which will hold the other part of the axis DC corresponding to the one fixed below, AB. The semi-armillary CVE, with radius CA or AV of 6 unciae, should be attached, to revolve round the pillar DB from C and E, and curving round it in an epicycle KLH in a gap KH three unciae wide. Half of this, VK, will therefore be $1^1/_2$ unciae. Within this space the globe of Venus can be fitted half an uncia in diameter, sufficient to show distinctly the markings of Venus on it. Allowing for the strips IN and LQ with axles IK and QH, more than half an uncia will be left, sufficient to allow freedom of movement.

---

[2] *1 Roman foot = 7.2 Imperial inches.* **P.F.**

The position of the Earth and its orbit round the Sun, in this system, can be correctly shown in the intervening space. For since the orbits of Venus and the Earth round the Sun are approximately in the ratio of 3 to 4, and the radius of Venus' orbit is fixed at 6 unciae, the radius of the Earth's orbit will be 8 unciae from A to Ω[3], and a semi-armillary GΩF turning round the axis of the Zodiac at G and F should be attached so that the Earth's globe Ω can revolve in its annual motion round the Sun A. If to the distance AV of 6 unciae is added the line VL of $1^{1}/_{2}$ unciae, there is only a half uncia left to the centre of the globe Ω. So if that globe were allotted its correct dimensions it would impinge in its movements on the half-epicycle KLH. Therefore the globe Ω can be made a little less than an uncia in diameter, since it is not necessary to show here the true size of the terrestrial globe relative to the planet Venus, which would have to be almost equal. But if it is wished to show the globes of the Earth and Venus equal, the radius of the circle AΩ will have to be a bit longer than 8 unciae, say $8^{1}/_{2}$. Likewise on that semi-armillary CΩF a half-epicycle must be fixed and strips placed to fit the axis of the Earth's daily rotation inclined $23^{1}/_{3}°$ to the plane of the Ecliptic. With these measurements a compact, three-dimensional armillary machine two feet in diameter BD will show the full range of Venus' phases and the positions of the markings as seen from the Earth at any given time.

Nothing else needs to be added to a full description of this machine, except a circle XSZTX a little over two feet in diameter should hold together the armillary machine ADYBA measuring two feet. This circle ZSXT should be set on the plane of the Ecliptic, and the axis of the Zodiac DGAB should meet it at right angles. That circle will be just like the horizon round globes on the armillary sphere, held up as is customary by supports $\Lambda\Gamma\Sigma\Pi$, and marked with the 12 signs of the Zodiac and the 360 degrees of the Ecliptic. Also, if it is wished, the months and even days of the civil year can be shown, as is usual on the horizon of globes; the situation of the spectator on Earth will be more quickly adjusted to the given date which corresponds to that position, if the semi-armillary FΩG is revolved around to fit the given day. Also the globe of Venus, V, will be placed in the correct position for that given day on this armillary machine more easily with the help of a table covering

---

[3] *Bianchini would have known that "A to Ω" (Alpha to Omega – cf. our "A to Z") would have reminded his readership of Revelations Chapter 1, line 8 where this phrase applied to the Deity meant that He was 'the first and the last' or 'the beginning and the end' of everything. Of all the Greek or Roman letters that were still available for Table V, his choice of Omega seems to suggest that he is 'nailing his colours to the mast' here and practically declaring his support for the Copernican sun-centred system; a senior cleric of his standing would be mindful of Galileo and hence play safe by sitting on the fence when it came publicly to choosing the Copernican theory; and the remarks of Dom J.F.Baldini about Bianchini's complete impartiality appear to have been written somewhat tongue in cheek.* **P.F.**

an eight-year period which we will publish below. For in the space of eight years Venus completes exactly 13 of its orbits round the Sun to within $1^1/_2$ degrees. There are in this eight-year period 2,922 days, and since Venus takes about $224^2/_3$ days to complete one revolution round the Sun, 13 revolutions will add up to 2,921, that is to say one day less than our eight civil years. In a day, Venus moves 1 degree 33 minutes in its orbit round the Sun.

And so, on this three-dimensional armillary machine (made according to the Copernican system because of the economy of construction) the positions of the Earth and Venus in relation to the Sun can be shown most readily. The explanation of the illumination will be provided through motion, when the lamp at A is lit, and the half-circle EHKC moved around carrying the planet Venus to the degree of the Ecliptic appropriate to its heliocentric movement at a given time, which must be ascertained from a table which we will provide. We suppose that the invention of Peregrino Mazza has been fitted to Venus' globe, to ensure that the axis of rotation is maintained parallel to a line from the Sun to the 20th degree of Leo and Aquarius. But if a device of the kind invented by Mazza cannot be acquired to ensure the parallelism of the globe's axis remains constant, this effect can be obtained quite easily by the use of a small magnetic compass.

The needle should be magnetised and secured in its case as is usual, then carried round the circle XSZT until the needle points to the centre of the circle, A, which is the Sun. For example, this would be at Z if the needle lay along the line ZY pointing to A. If at that point the needle happened to be at the 20th degree of Aquarius, the circle ZSXT should not be moved at all. But if some other degree of the Ecliptic happened to be at Z, the circle ZSXT should be moved round (raising the magnet off it, to prevent it from moving round with the circle) and when the point Z coincided with the 20th degree of Aquarius, the magnet should be placed on it pointing towards the Sun. The plane of the semi-armillary CVE which carries Venus and represents the plane of the axis of its revolution LN should also be pointed in the same direction as the magnetic needle. Then the armillary carrying Venus should be rotated along its orbit to some given date. Wherever the plane of the armillary happens to be, the parallelism of the axis can easily be adjusted by hand by placing the magnet on it. For since the magnetic needle, wherever it is taken, always by its own nature keeps to a line parallel to XY, this will act as a pointer, when placed by the globe of Venus, so that the axis LN can be arranged in the appropriate direction to be parallel (to XY). Therefore this defect in the machine can be supplied manually, the disposition of the magnetised needle being sufficient to determine the direction of the axis of rotation. When this is adjusted correctly, the illumination of the Sun, the phases of Venus and the appearance of the markings will be shown accurately through this machine, just as they really are in the sky.

A simpler machine that we have devised more economically can show all the phases so far mentioned just as well for one system as for the other. For an understanding of them requires nothing else to be displayed than the triangle formed by the centres of three globes, the Sun, Venus and the Earth. When a triangle of this kind has been constructed with the correct proportion of the sides and the sizes of the angles being kept, whether the one system or the other is chosen makes no difference, for the nature of the triangle is the same for observers following either system. Therefore the exact measurements of the sides and angles of this triangle will be supplied by this simplest of all machines, which I will now describe.

Let a rod BA be set up to represent half the axis of the Zodiac (Table V, Figure 1). The point A should be supplied with a lamp, as before, to represent the Sun. Around this rod or half-axis AB should be attached at E a quarter of the circle which carries round the globe of Venus, V, namely CHE, and as in the previous machine the epicycles KVH, IN and LQ should be added with their axles to show Venus' rotation and the parallelism of the axis. Then the circle of the Zodiac should be fixed on the outside just like the horizon on an armillary sphere, supported on four legs as in the previous machine, with enough space left for the arm EHV with its little circles HLK to carry Venus round the Sun. The width of this strip or Zodiacal band placed round like a horizon should be sufficient to allow the 360 degrees of the Ecliptic contained in 12 signs to be written on the inner part of the circle, and on the outside the 365 days of the civil year, equally distributed month by month, as is usually done on the horizon of an armillary sphere of a celestial or terrestrial globe. Finally the little globe representing the Earth should be fixed firmly to the end of a copper bar A (Table V, Figure 2), the other end of which, B, should be fitted with a device BC by which it can be attached to the circle of the Zodiac already placed round like a horizon and freely moved through the different degrees of that circle and placed on the individual days of the year marked on it, and the bar BA fastened securely there along with its globe by means of the screw D.

So with these provisions not only will all the phases be shown by this simple machine which were displayed with more difficulty in the previous ones adapted to both systems, but also they will be represented more accurately, no matter which system is chosen, for the addition of the device BC will enable the globe A representing the Earth to be moved forwards or backwards gradually from the Sun in proportion to the way that the distance varies between solar apogee and perigee, a facility which the previous machine did not possess.

Therefore let the line AV be the radius of Venus' orbit (Figure 1), taken to be the average distance of that planet from the Sun, as the ellipse which it describes is very near to a circle. We will divide this average distance AV into three equal parts. If to these three parts we add another equal one, we will have the Sun's average distance from the Earth, and the distance of the globe A fixed on the bar AB should be equal to this fourth part when attached by the screw D to the eighth degree of Aries and Libra on the Zodiac previously marked out and placed round as a horizon. The size of the device AB should be sufficient to allow it to move forwards inside the circle or backwards outside it as much as is required to show the difference in the Sun's distance at perigee and apogee compared with its average distance from the Earth. In either system the difference between minimum and maximum distance can be taken as about 1 part in 32, since the Sun's diameter at apogee appears to us as 31 minutes 38 seconds, and at perigee 32 minutes 44 seconds.

The position of Venus in the Zodiac as seen from the Sun for a given year and day should be sought from the eight-year table published below, and to the corresponding place on the circle showing the Zodiac on our machine, the globe V should be moved by its supporting arm EHV. From the given day of the year it will now be known in what degree of the Zodiac the Sun appears as seen from the Earth. Therefore the globe representing the Earth along with its rotating bar device described should be fitted to the degree of the Zodiac opposite where the Sun is seen to be, and in proportion to the distance from apogee, the bar should be moved forwards or backwards on the strip representing the circle of the Zodiac, and be held there firmly with the screw. For example, if it has to be placed for June 13, 1726, when the Sun will be in the 22nd degree of Gemini, and the Earth as seen from the Sun in the 22nd degree of Sagittarius, the bar B (Figure 2) should be placed at the point R (figure 1) of the circle ZSXT.

It is certain that the measurement of the sides and angles of the triangles formed by the three globes of the Sun, Venus and the Earth at a given date will be determined very accurately by this simpler machine, without showing any preference for one system over the other. For those who follow Tycho, imagine the Earth to be stationary, around which the Sun describes a yearly circle carrying Venus' orbit along with it, and taking round the whole triangle formed by these three bodies. The supporters of Copernicus, however, imagine that round the stationary Sun A, both Venus V in its orbit and the Earth in its ellipse travel forward towards that date in such a way that each one reaches the degree of the Zodiac as seen from the centre of the Sun, that this theory of their movements requires.

Therefore in this simplest of all machines and most accurate when the lamp representing the Sun is lit, the globe V representing Venus will reflect light in such a way that it will present the same phases to an observer situated at the globe representing the Earth, as the real Venus in the sky will present at a given day to telescopic observers on Earth.

Anyone who wishes, however, can demonstrate with larger globes everything this more compact one achieves with smaller ones, as long as the proportion of the sides and angles are the same as in this machine, however large a triangle is formed. At one corner of that triangle the globe of Venus with the markings depicted on it should be placed, and at another corner of the same triangle the lamp representing the Sun be made to direct its rays towards that globe by means of a magic lantern in a dark room, to a spectator situated at the third corner representing the Earth; the effect will be obvious without further explanation.

IX. If it is wished however to demonstrate also on a planisphere the same effects that we have shown on the three-dimensional machine, refer to Table IV. Let us suppose that the eye of the spectator is placed on the axis of the Zodiac at its north pole S, so that the point S is itself the line of the axis passing through the centre of the Zodiac and the Sun S, while the circle EFGH is the plane of the Ecliptic seen from its axis at right angles, and having its centre at the Sun, S. From that same centre of the Sun let Venus' orbit ABCD be marked, whose radius SB is in the ratio of 3 to 4 with the larger concentric circle's radius SF. The orbit EFGH will refer to the annual revolution of the Earth round the Sun, in the Copernican system, and the orbit ABCD will refer to Venus' eight-month path round the Sun in the same way. Divide the circle EFGH into 360 degrees, distributed among the twelve signs of the Ecliptic, so that we can measure Venus' movements by prolonging a line from the Sun through its globe right up to these divisions into degrees, and also reckon in degrees the progress of the terrestrial globe at a given date when the aspect of Venus' phases as seen from the Earth is being sought.

In order to calculate both these things more easily, we should remember what was said previously, namely that within a period of eight civil years Venus and the Earth in the Copernican system, or Venus and the Sun in the Tychonian system, return almost to the same points in their orbits that they occupied eight years previously. So let us take as a starting point the date in any epoch at which Venus is in inferior conjunction as seen from the Earth, that is when it lies at A on a line from the Sun S to the Earth E. From astronomical tables or ephemerides based on them, we know that in 1726 inferior conjunction occurred on April 6th, when both the Sun S and Venus A as seen from the Earth E were in the 17th degree of Aries, or (which is the same thing) as seen from the Sun, Venus and the Earth were in the 17th degree of Libra. Therefore the point F must be put in the 17th degree of

Capricorn, G in the 17th degree of Aries and H in the 17th degree of Cancer. So when the circle EFGH is divided into its 360 degrees, the beginnings of the signs Aries, Taurus, Cancer etc. can also be marked in their proper positions.

On the outer, larger circle EFGH, as well as the signs of the Zodiac the days of the civil year corresponding to them can also be marked, as is usual on the horizons of armillary spheres and celestial and terrestrial globes. This means that April 6th will be placed not on the 17th degree of Aries but opposite on the 17th degree of Libra, since the diagram of this planisphere is adapted in this position to the system of a moving Earth. A little later it will also be necessary to show how to assign the degrees if the planisphere is to be adapted to the system of a stable Earth.

Once we have chosen the epoch when Venus was in inferior conjunction on April 6th 1726, we will easily work out the positions of Venus and the Earth relative to the Sun for any day in the whole eight-year period. The situation of the Earth in its yearly orbit is given by the date itself, written on the circle alongside the degree of the Ecliptic which the Earth occupies as seen from the Sun. The position of Venus as seen from the Sun will be deduced from consulting a table which I provide. Anyone who wishes could work out this table by taking note of the following points.

Venus takes $224^{2}/_{3}$ days to complete its full orbit round the Sun, and $56^{1}/_{6}$ days for a quarter of it. We fixed the starting-point of that orbit at E, and the line of conjunction of Venus and Earth on April 6th, 1726, at the 17th degree of Libra as seen from the Sun. Therefore 56 days before April 6th 1726, that is to say February 9th, it was at the 17th degree of Cancer, at H. 56 days after April 6th, that is to say June 1st, it was at the 17th degree of Capricorn, at F, and another 56 days later, on July 27th, it was at the 17th degree of Aries. When it returned to E, where it had been on April 6th, $224^{2}/_{3}$ days later, the date was November 17th of that same year. But before that it had returned to H on September 22nd, namely $224^{2}/_{3}$ days after February 9th. So for the whole eight-year period the days should be counted, divided into groups of $224^{2}/_{3}$ days, and placed on the table in four columns of numbers, ABCD, indicating the days and months at which, as seen from the Sun, Venus is in the 17th degree of Libra, Capricorn, Aries and Cancer. Other columns of numbers should be placed alongside these first four, indicating the day of the year and month when Venus, as seen from the Sun, first enters each sign. This is easily determined by working out the amount of time taken by Venus to move through 13, 17, and 30 degrees. It travels through 13 degrees in 8 days. Therefore alongside columns ABCD let us place another four, following them with dates eight days later, which will show the day of the year when Venus, as seen from the Sun, enters the signs of

Scorpio, Aquarius, Taurus and Leo; and four other columns preceding them, showing dates 11 days earlier (the time taken by Venus to travel 17 degrees of its orbit). Thus we will cover all the period of that eight-year span, when Venus as seen from the Sun entered the signs of Libra, Capricorn, Aries and Cancer. Finally two more columns of numbers should be introduced among these, moving forward the dates by 19 days (the time taken by Venus to traverse a sign) and the table will be complete. It will show for a full eight-year period the ephemeris of Venus' heliocentric motion when it enters the various signs, and at the 17th degree of the cardinal signs of Libra, Capricorn, Aries and Cancer, in the first of which the Sun and Venus were in conjunction as seen from the Earth on April 6th, 1726, in the same plane as the circle through the poles of the Zodiac and the 17th degree of Aries.

At the end of this eight-year period Venus will be in conjunction with the Sun one day earlier as seen from the Earth in a similar plane through the poles of the Ecliptic and the 16th degree of Aries. Therefore during the following eight-year period from April 5th 1734 to April 4th 1742, the same table will show Venus' movements as seen from the Sun relative to the points of the Ecliptic given above for a certain day of the year, by taking away one day from the date given for the previous eight years. As each eight-year period ends one day earlier than the preceding one, this one table will give the ephemeris of Venus' movement sufficiently accurately for us to observe its illumination for eight or ten preceding eight-year periods and the same number following, if for every preceding eight-year period we add one day, and for every following eight-year period we subtract one day.

For example, let us enquire the location of Venus in the Ecliptic on October 14th, 1666, while it was being observed by Cassini as shown in the French *'Journal des Scavans'* published in 1667, page 257, and Ozanam's *'The Celestial Sphere'*, page 80. Eight of the eight-year periods will have elapsed from October 14th 1666 up to October 14th 1730. On October 14th 1730 Venus will be in the 25th degree of Aquarius as seen from the Sun, according to our Table, since it will be entering Pisces on October 17th, and in three days from October 14th to 17th, Venus moves through five degrees in its orbit. Therefore in 1666 Venus was in the 25th degree of Aquarius eight days after October 14th, as seen from the Sun, and eight days earlier, that is October 14th 1666, it was in the 12th degree of Aquarius, since it traverses 13 degrees in its orbit during that time.

On the same day of that year and month the Sun, as seen from the Earth, was in the 21st degree of Libra, but the Earth as seen from the Sun was in the 21st degree of Aries. Therefore two lines from the Sun, one to the Earth and one to Venus, would form an angle of 69°. If this angle is known of a triangle formed by lines from the centres of the three globes, the Sun, Venus and the Earth, the rest can also be worked out, since the two lines from the Sun's centre are radii of the two orbits, Venus round the Sun and the Sun round the Earth in the Tychonian system, or Earth round the Sun in the Copernican system (which will be used to show the phase of Venus, as it will serve equally well for either system), and these radii are approximately in the ratio of 3 to 4 or 59 to 81, as we have already stated. Therefore the whole question of the illumination of Venus and what proportion of the lit hemisphere was presented to us that day can be worked out from the triangle.

However, it will be seen more readily in the armillary machine simply by placing the semi-armillary carrying Venus at the 12th degree of Aquarius, and the one representing the Earth at the 21st degree of Aries. When the lamp representing the Sun is lit in the middle of the machine, it will illuminate the whole of the hemisphere of Venus turned toward it. More than half of this hemisphere will be turned towards the eyes of the observer on a line from the Sun through the 21st degree of Aries to the place occupied by the terrestrial globe. We will see Venus therefore like the gibbous Moon, just as shown in Cassini's observation.

Let us take as another example Cassini's observation of April 28th of the following year, 1667. Since it was eight of the eight-year periods earlier than April 28th 1731, Venus as seen from the Sun was at the same degree of the ecliptic on April 28th, 1667, that it will be on May 6th, 1731, that is the 27th degree of Sagittarius. Therefore on 28th April 1667 Venus must have been in the 27th degree of Sagittarius as seen from the Sun, while the Earth as seen from the Sun was in the 9th degree of Taurus, and lines from the Sun to Venus and Earth formed an angle of 50°. So from the known ratio of the radii of the paths of Venus and the Sun, or of orbits of eight months and a year[4], of 3 to 4, the complete shape and measurement of the triangle can be worked out from the three lines, from Sun to Venus, from Venus to Earth and from the Earth to the Sun, and the amount of the illuminated hemisphere of Venus presented to us is revealed. But leaving aside the armillary machine, let us return to the planisphere. We said that with the help of the table of eight-year periods we have produced, the position of Venus on the Ecliptic as seen from the Sun in its daily orbital motion could be ascertained.

On the diagram of the planisphere already shown, therefore, in the plane of the Ecliptic (Table 4, Figure 1), the spectator's eye is placed at the pole of the Ecliptic, where it looks down perpendicularly on the centre of the Sun,

---

[4] *Bianchini here seems to be about to allude to Kepler's third law: $Period^2$ is proportional to $Mean\ Radius^3$, but has forgotten to explain its relevance here; alternatively, this might have been the result of a hasty deletion.* **P.F.**

S. From this centre S two straight lines or two threads should be extended outwards, one to the place in its orbit occupied by Venus according to that Table, the other to the place of the terrestrial globe according to the system of a moving Earth (or to the place opposite the Sun in the system of a stable Earth, which has the same effect) to establish the angle which these two lines or radii form at the centre of their circle, S.

Having done this, the beginnings of the Signs should be marked on the outer circle of the annual orbit EFGH on which 360 degrees have already been shown. As it was the point E, to which a straight line from the Sun S had passed through Venus on April 6th 1726 at A in conjunction with the Sun at the 17th degree of Aries, so that the spectator situated at E would see Venus A and the Sun S along the line of sight EASG together, the point G of the Earth's annual orbit will be assigned to the 17th degree of Aries, and its opposite point E to the 17th degree of Libra, where Venus and Earth are in conjunction as seen from the Sun. The point F, a quarter of a circle away from E, will be fixed at the 17th degree of Capricorn, and opposite it H at the 17th degree of Cancer. From H, 13 degrees further towards E will complete the 30 degrees of Cancer, and there the start of the sign of Leo must be marked, and similarly all the other signs of the Zodiac must follow in their correct order.

On the same circle EFGH an outer, concentric strip should contain the calendar of the civil year divided into its 365 days, as is usual on the horizon of an armillary sphere and a celestial or terrestrial globe, assigning the date April 6th to a point E at which conjunction of the Sun and Venus happened in the year 1726.

Thus when a planisphere is divided into degrees, it is easy to assign both Venus and the Earth to their correct positions in their orbits for a given day. The Earth's position will indeed be apparent in this system from the calendar itself, while that of Venus will be found in the Table of eight-year periods already explained. To understand the illumination of Venus' globe as seen from the Earth, it is sufficient to show on the planisphere a cross-section of that globe lying on the plane of the Ecliptic, as we have done at points A, B, C and D. On these the hemisphere turned towards the Sun is shown light and the other hemisphere dark. How much of the illuminated part is turned towards us will be perceived by the following method, which will be better understood by using an example. The section of the hemisphere lit by the Sun and observed on Earth on February 9th, 1726 should be sought. From the Table of the eight-year motion of Venus, it is established that Venus was at D in the 17th degree of Cancer. The position of the Earth was, however, at the 18th degree of Leo at Σ, and the Sun was seen in the opposite direction at the 18th degree of Aquarius along the line ΣSV. Venus as seen from the Sun on the line SD was lit in the hemisphere LXKI, and at that distance from the Sun it can be taken as being bounded by the plane IL perpendicular to the line SD. That same Venus is seen from the Earth Σ along the line ΣD, which is the axis of the hemisphere turned towards the Earth and cuts the dark hemisphere at f. From that axis ΣfD a quadrant of the circle, fr, is cut on one side, and another one, fp, on the other side, to show the hemisphere rfp turned towards the Earth. If the Earth were to be found on the extension of the straight line IDL, half of the hemisphere of Venus lit by the Sun would be visible to Earth and half of the dark one, and the shape of Venus as seen from the Earth in that plane would be like the half-moon at quadrature. But when Venus is seen from the Earth at Σ through the line ΣfD, the whole of the arc fL from the dark hemisphere is turned towards the Earth as well as a complete quadrant, and from the illuminated quadrant an arc equal to fL itself is taken away from the observer watching Venus from the Earth. So by counting the degrees of the arc fL, the shape of the crescent of Venus can be shown with that segment subtracted from the half-phase or dichotomy. To measure this in practice on the planisphere, it will suffice to affix a strip of paper like a rule, rotatable about the point S, and at the point D of this strip of paper to place a circle LXKIML likewise fixed by its centre and rotatable about the point D. The circumference of this little circle, divided by its own scale into 360 degrees, will always show the zero of its graduations on the axis of illumination, which is the line from the centre of Venus to the Sun's centre. In the example given for February 9th, the start of the numbering of the 360 degrees is set on the line DS at X. The circumference of the circle XIMLX will show at the arc Xf by how many degrees the arc Xf exceeds the quadrant XL. This is the amount to take away from the quadrant, to display the defect in illumination of the crescent Venus at this position.

In order to draw the shape of Venus' disk as it appeared to us on that day, the spectator placed on the plane of the Ecliptic at Σ should imagine the plane at right angles to the Ecliptic at D, the centre of Venus' globe. The line of sight ΣD from the point Σ to D will be represented in Figure 2, Table IV, as the point R, and the plane of the Ecliptic as the line HRS. The plane at right angles to this through the centre R will be shown by the line IRP. At the great distance of Venus from us, the arc designated by fL on the planisphere can be represented as curving from R to G in this Figure 2 or, which is the same thing, by measuring its degrees on the circumference of the circle in this diagram from P to L and from I to N and joining LN which will cut the line HRS at G. The terminator IGP will show the crescent figure of Venus as it appeared to us on that day, smaller than at dichotomy by the segment IGPR.

If it is desired to reach the same conclusion without the help of planisphere, paper and little circle, this can be achieved by calculation with the help of the Ephemerides alone, as follows.

On the diagram of the planisphere (Table IV, Figure 1) let us consider the triangle SDΣ, whose angles can be calculated from consulting our table of eight-year periods and the Ephemerides. The angle DSΣ is the difference in longitude of Venus and the Earth as seen from the Sun, and can be shown to be 33° from the Table of eight-year periods on that day, since Venus was in the 16th degree of Cancer and the Earth in the 19th degree of Leo, in that with respect to the Earth, the Sun was in the 19th degree of Aquarius. The Ephemerides, which confirm that the Sun as seen from the Earth was in the 19th degree of Aquarius on that day, also show Venus as seen from the Earth in the 7th degree of Aries. Therefore the angle DΣ S is 48°, and the angle SDΣ can be worked out as 99°, because the angles of a triangle add up to two right angles. Subtract from this the whole quadrant XL of 90°, that is to say half of the illuminated hemisphere on the axis SXD, and an arc of 9 degrees remains. This is the amount of its dark hemisphere, in addition to a complete quadrant, which Venus displays to us as seen from Σ. In Table IV, Figure 2, the same arc is represented by the line RG, and the crescent phase of Venus by IGLOI.

In Cassini's observation of April 28th 1667, we saw that the angle was 48° formed by the two lines from the centre of the Sun, one to Venus which according to our Table was in the 27th degree of Sagittarius, the other to the Earth in the 9th degree of Scorpio (opposite to the 9th degree of Taurus in which we then saw the Sun). From the Ephemerides, it is established that lines from the Earth to the Sun and to Venus formed an angle of 45°. Adding this to the 48° of the previous angle gives a total of 93°; therefore 87° remains from the sum of two right angles as the value of the third angle formed by lines from the centre of the Sun to Venus and the Earth. So Venus' disk was only 3 degrees short of perfect dichotomy as seen by us, as Cassini showed in his observation of that day in which it is described as 'almost half'.

I think we have explained adequately the easy method of ascertaining the phase of Venus day by day during the eight-year period, both on the three-dimensional armillary machine and on the planisphere. It remains for me to show on the same planisphere the parallelism of Venus' axis of revolution. This can be done efficiently on the planisphere. For when we observed it, the plane of this axis of revolution at right angles to the Ecliptic passed through the Sun when Venus was at the point of its orbit around the 20th degree of Leo and Aquarius, namely at points R and V on the diagram of the planisphere (Table IV, Figure 1). It follows, therefore, that the line MRKSV marks the plane where the axes of revolution and illumination join.

At whatever point of its orbit Venus may be, there let a line parallel to MRKS be drawn through its centre. It will be understood that the plane raised through this line perpendicular to the Ecliptic will be the same plane in which Venus' poles of rotation lie, the North one raised 15° above the Ecliptic and the South one depressed by an equal amount below, as was shown adequately on the armillary machine's construction. The form of that planisphere, which is more compact and takes up less space when using the diagram of the Copernican system, can also be made to apply to the Tychonian system, but needs more space to show double the expanse of the former, to be adequate to show the globe of Venus to the same scale. However, I will explain the method of its construction to assure everyone that these phases of Venus can be shown equally well in both systems.

The planisphere therefore will follow the Tychonian system if the Sun's distance from the Earth SE is kept, and from the centre E a circle with radius ES is described to show the Sun's annual motion round the Earth. A sixth of this is shown here by the arc marked φΩS. A revolving rule should be placed along SE with one of its ends fixed firmly as an axle at E representing the Earth, while the other end representing the Sun, S, should be moved round along the line of the letters φΩS and carry with it in the manner of an Epicycle moving from its deferent the orbit of Venus ABCD, keeping the Sun always at its centre. The circle of the Sun's orbit round the Earth φΩS etc. should be divided into its own 360 degrees and 12 Signs, just like Venus' orbit round the Sun. After doing this we should note from the Ephemerides that on April 6th 1726 the Sun S and Venus A were seen from the Earth in the same plane through both their centres to G at the 17th degree of Aries on the Ecliptic, so the point A of Venus' orbit seen from the Earth will also coincide with the 17th degree of Aries. The progress of Venus round the Sun should be according to the order of the signs from A to B, C and D, and also the progress of the Sun in its annual orbit round the Earth should be according to the sequence of letters φΩS, which is an arc showing a sixth part of the whole circle, and half of that sextant ΩS or φΩ is equal to one sign of the Zodiac, divided into 30 degrees as is customary. From this sextant of the circle, SΩφ, the remaining parts of it can easily be imagined around the centre E with radius SE or Eφ . We have not shown it here, to avoid presenting a diagram of inconvenient size for the reader and printer of the book.

Let us imagine on the planisphere EFGH lines parallel to the diameter EG. Although the diameter EG of our planisphere includes the radius of a circle described from that same centre E and having the same centre as the Sun's orbit φΩS, and the 17th degree of Aries is shown by the direction of the line ESG, in the plane of which, at right angles to the Ecliptic we said that Venus and the Sun as seen from the Earth E were in conjunction on April 6th 1726, because of the very great distance of the Zodiac from the Earth and the Sun, from where the whole expanse of the Sun's annual orbit round the Earth can be regarded as a single point, lines parallel to the diameter EG or the radius ES can all be regarded as pointing to the same 17th degree of Aries.

Therefore when the rule ES has been made to rotate about its centre E through the Sun's annual orbit φΩS, and when the circle ABCD representing the orbit of Venus round the Sun has likewise been carried along together with the end S of the rule, the point C of this circle of the orbit of Venus will always have to be so arranged that, whilst the Sun's centre S is moving around whether at Ω or at φ or at any point of the circle, the radius SC should fall on one of the parallels to the line ESG that are marked on the planisphere; so that it would always reach towards the 17th degree of Aries in the Zodiac, to where it should point.

At the centre S of the same orbit of Venus an alidade[5], or a moveable rule RS should be fixed at the same point S, and rotatable about that point and at its end R it should carry around the little circle KIML always fixed to the point R, and the circumference of this little circle should be divided into 360 degrees, starting at the point K, which a line drawn from the centre of the Sun to the centre of Venus, R, (for example, SR on February 9th, 1726) shows to be the pole of the hemisphere lit by the Sun, IRLK.

With these small modifications to our planisphere it is made capable of showing the phases of Venus in the Tychonian system. For at a given day of the year and month, when the Sun's place in the Ecliptic is known, whose twelve signs are marked in degrees on the orbit φΩS, one end S of the alidade ES holding the orbit of Venus ABCD attached to it will have to be fixed at that place. Then the orbit of Venus must be rotated around the centre S until the radius SC falls on one of the parallels to the line EG on the planisphere pointing to the 17th degree of Aries and fixed there. Then our Table of eight-year periods of Venus' motion round the Sun should be consulted to find the place in the Zodiac where Venus appeared on that day as seen from the Sun. At that degree marked on the circumference ABCD the alidade should be turned round with the small circle fixed on the end representing Venus. To us watching from the Earth E it will make the phase clear if a string from

---

[5] *Alidade - A rule and an alidade both form parts of an astrolabe. The rule is a rotatable pointer fixed to the centre of the astrolabe and marked off in degrees of declination for use with a star map below it, whilst the alidade resembles the rule, except that it has sights, and is used in measuring altitudes when the astrolabe is suspended vertically, and is on the back of the astrolabe whereas the rule is on the front.* **P.F.**

the centre of E is extended to the centre of Venus' globe. For this thread will cut the circumference of Venus, on which 360 degrees should be counted from the point opposite the Sun where the axis of the illuminated hemisphere lies. The thread extended from E will show how many degrees lie between our axis of vision and the terminator of Venus; this was our aim.

It seemed right to me to include also in the construction of the planisphere this method of showing that in the Tychonian system the phases of Venus are the same as in the Copernican, to stress that these phases did not disprove either system. But the reason I chose to publish the diagram of the planisphere for the system of a moving Earth was firstly so that it would take less room and fit the size of the paper used by the printers, and secondly because it would be easier to work out the movements and quicker also to understand them, in that a single glance can clearly show the two orbits and their effects without the necessity of employing rulers and little circles rotatable about moveable centres, and systems of numbering which have to be adapted to parallel lines; all of which we warned would have to be employed in the diagram showing the Tychonian system.

X. I have said that both the three dimensional armillary machine and the planisphere that I have described here will serve to give a practical demonstration to spectators of the phases of Venus and the parts of its illuminated hemisphere visible from the Earth, and in particular to show the progress of the markings and the parallel circles they describe round the axis of rotation. In the case of the three-dimensional armillary machine first described there should be no difficulty at all, for the solid globe of Venus is represented by a small solid globe lit by the rays of a lamp like the Sun on the same machine. On the planisphere, however, where the three-dimensional aspect of solid bodies is distorted by being shown in a two-dimensional plane, the representation is much more difficult, and it is much harder for the faculty of imagination to perceive the true nature of what is being demonstrated. Therefore I would advocate that whenever it is desired to use the planisphere to give a visual demonstration of the phases of Venus and its markings as seen from Earth, as well as the planisphere already described, a small globe should be used, at least $1^{1}/_{2}$ unciae in diameter, with the markings observed on Venus depicted upon it (the method of showing these will be given in the next Chapter) and placed on a stand half a palm high with a semicircular arm to be attached to each end of the globe's axis at the poles, permitting the globe to turn round with an inclination of $15°$ above and below the plane of the planisphere. Another stand of the same height, half a palm, should be placed on the centre of the planisphere and provided with a lamp to play the role of the Sun in illuminating Venus' globe. At a given day of the month and year the positions of Venus and the Earth should be sought by the method shown above, using the eight-year Table produced previously. Over the place in the planisphere where Venus should be, the little globe should be placed, supported on the half-palm stand, and its axis of rotation should be turned towards a direction parallel to the diameter of the planisphere, passing through the 20th degrees of Leo and Aquarius. When the lamp in the centre of the planisphere is lit, it will light the little globe in the same way that the Sun's rays are lighting the real Venus on that day. If the globe is turned on its axis, the movement of all the markings and parallels described by them will be observed, as long as the eye of the spectator is placed on the plane passing through the centres of Venus' globe and the Earth, situated in the correct relative positions in the planisphere according to both their orbits round the Sun in that system.

CHAPTER IV

THE CELIDOGRAPHY, OR DESCRIPTION OF THE MARKINGS OBSERVED ON VENUS' GLOBE, IS PRESENTED AND NAMES ARE ALLOTTED TO THEIR MOST IMPORTANT PARTS.

Summary of the Chapter

I. *In order to observe all the markings on the planet Venus, it is necessary to wait for the times when they are turned towards the Sun and to us, and nearer to the Earth. To map them, a place must be chosen which is correctly situated both as regards the observer and the axis of rotation. For either the spectator should be situated outside the globe of the Earth and placed on the axis of the Ecliptic; or located in the plane of the Ecliptic upon the Earth's globe where we are.*

II. *In either position the lines described by the markings should be considered as they rotate round the axis of revolution, so that the appearance presented to the eye of the spectator can be understood whether in the form of ellipses, circles or straight lines.*

III. *By applying this theory to the observations of the Venusian markings made in 1726, it was established that the plane through the axis of rotation and the Sun's centre cut the Ecliptic at the beginning of March 1726 about the 20th degree of Leo and Aquarius.*

IV. *When the markings were thus identified, and also their axis of rotation (the inclination of which, 15 or 20 degrees above the plane of the Ecliptic, was deduced from this), armillary machines can be constructed, both of the three-dimensional kind and planispheres, and maps to show the situation of the markings on the globe of Venus and their rotation and revolution day by day.*

V. *Two different types of sheet maps are described, such as those used by geographers and those who map the seas.*

VI. *For the sake of clearer understanding it, it is better to start with the parallel maps.*

VII. *Seven notable markings near the equator of Venus and two near the poles should be given the names of 'maria', just as on the Moon.*

VIII. *Names are assigned to each.*

IX. *Another arrangement is proposed of a map made in the form of circles, as geographers use a planisphere to represent the Earth's globe.*

X. *A solid globe with the same markings reveals all the phases more clearly.*

XI. *The method of constructing this and fitting it on to the armillary machine to imitate the Sun's light accurately to show the markings is explained.*

I. From what has been said so far, it should be obvious that all the markings on the globe of Venus can be observed and described by us if we wait for the times when that part of the planet's globe to be described and observed is lit by the Sun and also turned towards ourselves, the observers; and also when Venus is near enough to the Earth that with a telescope magnifying 100 times, the surface of Venus is equally well seen through the optic tube as the lunar markings commonly called 'maria' are with the naked eye at the Moon's average distance from the Earth.

To find out what parts of Venus' globe will be turned toward the Sun and us at that distance from the Earth, and the suitable times for observation, look again at the diagram of the planisphere (Table IV) and consider Venus' globe at that point of its orbit where the axis of revolution lies on the plane through the centres of Venus and the Sun, which we have said occurs when Venus is at about the 20th degree of Leo and Aquarius as seen from the Sun. It was seen from the Sun in the 20th degree of Leo in Table IV Figure 1 at R on March 1st, 1726, when the plane through the axis KRM prolonged beyond K cuts the centre of the Sun S, and prolonged beyond M cuts the Ecliptic at Σ, the 20th degree of Leo. If the extreme points of the axis were K and M, which actually lie on the plane of the Ecliptic, the axis of revolution KM would point directly at the Sun's centre. Its rays would strike the pole K of that axis at right angles, and would light as far as the great circle IL on the globe of Venus (or there would be no noticeable difference from a great circle because of the Sun's great distance which would hardly give an appreciable parallax). This means that IRL, the terminator dividing the lit hemisphere RLKI from the dark one IRLM, would be identical with what we call the Equator, the circle situated an equal distance from both poles of revolution, K and M. But since the pole of revolution in the hemisphere IRLK is raised above the plane of the Ecliptic and point K by about 15 degrees, we have said that it should be represented by the point Z (according to the principles of celestial perspective mentioned in Chapter 3, Section 9). This point is determined by marking the arcs Ka and Kb fifteen degrees on each side, and joining them by a line ab which cuts the axis at Z. And since that pole Z raised thus above the Ecliptic is found in the hemisphere of the stellar sphere containing the North Pole of the Zodiac, we will call Z the North Pole of Venus. The pole g, lying in the opposite hemisphere of the sky, which is depressed 15 degrees below the plane of the Ecliptic, will be called the South Pole. The Equator, or great circle 90 degrees from both the North and South Poles, will be ITL, and on that diagram will be shown in accordance with the rules of perspective already explained; that is to say a 15 degree arc TR equal to ZK should be measured from R to T and an ellipse ITL drawn, the long and short axes of which RL and RT are now established. But the lesser circles described by the Venusian markings as they rotate round the axis ZRg, which are parallel to the Equator, will have to be represented in a similar way by ellipses parallel to this one, for example nRX, etc.

This, therefore, is the explanation of the perspective obtained by a spectator positioned at the north pole of the Ecliptic looking down from there on the globe of Venus orbiting the Sun as it reaches the point R, where the plane through the axis of rotation MRK passes through the Sun and the 20th degree of Leo and Aquarius. But to the eye of the spectator situated not at the pole but in the plane of the Ecliptic, the parallels described by the markings on Venus' globe will present a different appearance. For a marking on Venus' equator, which as it revolves will show a spectator at the pole of the Ecliptic a half-ellipse ITL in the plane of the northern hemisphere, will show to another spectator in the plane of the ecliptic at the point Ψ, where the diameter of Venus IRL at right angles to the axis of rotation cuts the Ecliptic, the great circle of Venus' equator ITL as a straight line. For the spectator at Ψ is in the same plane, whereas the spectator outside that plane, at the pole of the Ecliptic, would see that circle as an ellipse, according to the rules of conic sections. But the circles described by the other markings on Venus' disk will be more like straight lines, the nearer they are to the planet's equator, to the spectator situated at Ψ.

II. By applying this theory to our observations of the progress of the markings appearing on Venus' disk from February 9th to March 5th, we discovered that it was on the days near March 1st that the parallels described by the markings as they rotated were nearest to straight lines. Therefore, it was necessary to deduce that the plane of Venus' equator at that time very nearly passed through our line of sight as we were observing from the plane of the ecliptic.

In order to determine the point of the Ecliptic more precisely where Venus' axis of rotation is in line with the Sun, and where the plane of Venus' equator meets the Earth, let us divide the quadrant of Venus' orbit DRA into 5 equal parts. As it takes 56 days to traverse the whole quadrant DA, it will complete a fifth of it, DΔ, in $11^{1}/_{5}$ days. The date February 9th had been marked at D, where Venus then was. So the date February 20th will have to be marked at Δ, where it will then be; just below R the date March 3rd, and at the next fifths the dates March 14th and 26th, so that it will reach point A on April 6th, when the Sun, Venus and the Earth are in the same plane SAE from the centre of the Ecliptic S produced to the 17th degree of Libra, SAE. On the edge adjoining the circle of the Earth's orbit HΣE, the days of the month have likewise been marked to show the daily position of the Earth in this system.

From the centre of Venus' globe at R, therefore, on March 3rd, when the axis of rotation RZ lies on the plane ΣRS passing through the Sun, let a line RΨ, which will be in the plane of the Ecliptic, be drawn perpendicular to this plane ΣRS. If the Earth and the observers of Venus upon it were at Ψ, they would see the plane of Venus' equator ITL extending in a straight line and the latitudes parallel to its equator diverging only slightly from straight lines when the markings rotate along with Venus around the axis gZ, that is to say the circles nRx etc. In 1729 about March 27th, Venus will return to R, and the position of the Earth in this system will approximate to the point Ψ, as can be seen from our eight-year Table; and in the Tychonian system the same triangle RψS will be formed by the motion of the Sun carrying Venus with it around the stable Earth. Therefore, that will be a very suitable time for repeating this experiment and observing the progress of the markings in their parallels along almost straight lines on Venus' disk as seen by us then. But because on March 3rd 1726 when Venus was at R the Earth's position was not at ψ, the 8th degree of Libra, but at Γ, the 13th degree of Virgo (and in the Tychonian system with the Sun moving and the Earth stationary the same triangle RΓS was formed), the Equator of Venus ITL and the parallel circles described by the markings nRx etc. appeared to us in the form of ellipses of narrow width which could be deduced from the arcs Lu and ZK. This theory, compared with the diagrams of observations, clearly proves that the line of Venus' axis of rotation passes through the Sun around the 20th degree of Leo in Venus' orbit, and its opposite point, the 20th degree of Aquarius, and this point will be defined more accurately by those who complete the observations of the rotation of Venus' markings in March and the beginning of April 1729.[1]

Meanwhile, as the positions of the poles and equator of Venus were established sufficiently accurately from the 1726 observations, it was not difficult from them to fix the location of each marking and show it both on the globe and the map, just as geographers are accustomed to do, when they fix what is called the latitude of the Earth's seas and continents by their distance from the Equator, and what is called their longitude by other circles at right angles to those of latitude and intersecting at the pole, and then mark them both on maps and planispheres and on a solid globe.

III. It was not possible, however, for the whole globe of Venus with the markings on both North and South hemispheres to be mapped from the observations from February 9th to April 6th, while Venus was traversing the quadrant DRA of its orbit round the Sun, and the Earth was then situ-

---

[1] *This would be the year following the publication of this book: evidently Bianchini believed that he had not long to live and was in a hurry to publish, so that others could complete his work.* **P.F.**

ated within the arc ΣΠΨE, because although Venus in that situation displayed all its markings to us, by revolving in 24 days, except a few near the North Pole z, the whole hemisphere IeMduL was not lit by the Sun and could arouse no sensation in our eyes of itself or the markings upon it. So while Venus was at R on March 1st, whatever markings lay in the North hemisphere on the arc ZR at 75 degrees (since ZK and RT occupy 15 degrees) were exposed to the gaze of watchers at Ψ for 12 days and even to watchers at Γ. Those which were actually on the equator of Venus while traversing the arc TL were not lit by the Sun, and so did not betray their presence to watchers between Γ and Ψ, but once they had passed the point L in the next quadrant of their revolution they were rendered conspicuous by the Sun's rays now falling upon them. Finally those which lay beyond the Equator ITL towards the South Pole for another 15 degrees or more of latitude received no ray of solar illumination to reveal them to us. But when Venus had crossed into the quadrant of its orbit AB, where it was from April 6th for the next 56 days up to June 1st, when it reached B, while the Earth was between E and F, the southern hemisphere of Venus was then lit by the Sun both at and beyond the equator, and revealed to us watching from Earth the markings which had previously been hidden; and several others which we had specially noticed from Σ (on both sides of the Equator within the arc LQ to the South and an equal distance North from N to I) were again clearly seen. Thus a chance to map the whole globe of Venus was offered to us, part of it in February and March, and part in April and May. After June 1st, however, it became so distant from the Earth that the markings, though turned towards us and lit by the Sun, made only vague impressions on our eyes, even with the large telescope of 90 or 100 palms that we were using. But as they had appeared clearly enough in previous observations, we were able to give a complete description of the globe (with the exception of only a small part near the North Pole), and make a map and also a globe following geographical and astronomical conventions with circles of latitude and longitude. Finally we uncovered even this North polar region of Venus when lit by the Sun, and charted it in July and August 1727. So at last we have seen the whole surface of that planet, and now present it to the public.

IV. We have tried, therefore, to follow the practice of astronomers and geographers when showing the constellation patterns and the configurations of lands and seas on a globe, by displaying both on maps and globes the markings observed on Venus in this two-year period, with an accurate representation of their relative sizes and shapes as far as possible. It is appropriate to mention the method and advantages of each means of representation, to justify our reasons for adopting both models to serve our needs.

V. As regards maps, geographers commonly use two quite different types.

Some maps portray sections of the globe as it would appear to a spectator situated above its surface and looking down directly at a particular section exactly as the eye sees it. There are planispheres of the Earth like this which are made according to the rules of perspective in a flat circle with the equator shown as a straight line at right angles to a certain line of Meridian, like a canvas in front of a painter at a point on the globe's surface from which a line straight to the centre of the globe forms the axis of that meridian. The distortion that this produces has some advantages, but, as on the painter's canvas, equal portions of the globe are at unequal distances from the eye so that in consequence the more remote areas are compressed while the nearer ones retain their true dimensions. This means of representation is not ideally suited to our mapping of the Venusian markings. For since that geographical section shows areas over the surface of the globe that appear more and more distant to the eye, and for that reason depicts them as smaller and smaller, as we look at the globe of Venus, our line of sight meets the protruding curve of the globe at the central point turned towards us, and that portion appears larger than others which are equal but are turned away more from our line of sight. The distortion produced by this geographical section, therefore, and the projection commonly used when showing the curved globe of the Earth by hemispheres represented on a flat map by circles, does not seem appropriate to use here, where we seek to preserve the true proportions of the object more in keeping with the angles of curvature of the hemisphere which we see in our actual observations of Venus.

It would be preferable to show those markings by means of segments that could be glued to the surface of a globe, as is customary on artificial terrestrial and celestial globes. Certainly this means of representation is the best of all, but it comes under the heading of globes rather than maps. We will show below its application to a solid globe representing Venus. The sections to be fitted to such a globe, when shown on a flat piece of paper greatly disrupt the continuity of the drawing, and fail to show how the markings on one section link with those on the next. Before starting on a description of this method of attaching sections to a globe, therefore, I think it will be better to copy the kind of map devised by those who chart the oceans to show the Venusian markings on a flat surface.

That type of map shows all the lines of longitude as straight and parallel, not curved and joining at the poles. The Equator is represented by a straight line cutting these meridian lines at right angles, and so are the lines of latitude. Though they are circles parallel to the Equator, they are shown as straight lines at right angles to the meridians. Sailors like to use these maps to link the wind-flow patterns more easily to the various lines of longitude. But admittedly this means of representation expands the polar regions more than is correct, but this hardly deters sailors from preferring their use. For as regards navigation, the most frequented regions of the Earth are the tropic and temperate zones (for there is hardly any profit in navigating the polar regions, and furthermore this is hindered by ice and by nights which last for almost six months). The foreshortening that occurs progressively in the parts of the globe from the Equator through both temperate zones is not retained on these maps. On the other hand, sailors soon learn the facts, which can be added to the charts, concerning the amount of miles or yards by which one degree at the Equator surpasses the degrees in the parallels of latitude which follow in sequence. This distortion does not much alter the shape of the regions in the tropical zone, and up to the middle of each temperate zone, and their continuity can be seen over the whole of the globe.

VI. In my description of the Venusian markings, and the way they interconnect throughout the globe's surface, I prefer to use this latter type of map, persuaded by the same considerations which recommend maps of the second type to those who sail the seas. For the markings around Venus' equator to about 60 degrees latitude on either side of it are the ones which can be observed more clearly but the remainder near the planet's poles either are not turned towards us when lit by the Sun, or else are very far distant when either of the poles is turned towards us and the Sun at the same time, and do not give us an equally clear picture of their shape as those nearer the Equator. In these latter the separate bays, promontories and curves can be identified so distinctly in these long telescopes when Venus is near the Earth, that names can be assigned to the individual bends, curves and promontories, thus enabling them to be recognised again as the planet revolves, when after a full rotation of 24 days they return to show the same aspect to us. Observational experience shows us that recognition of the markings will proceed with greater ease and accuracy if we start our description with those situated near Venus' equator, plotting their positions according to their latitude and longitude on the map ruled out with lines intersecting at right angles (Table VI, Figure 1).

Therefore let the line $+Z$ represent the plane of Venus' equator, that is to say the greatest circle on its globe, equidistant from the poles of rotation, which if produced through the centre of the globe will meet the axis of rotation at right angles. Then let other lines be drawn perpendicular to this line $+Z$ and parallel to each other, RS, $\Gamma\Delta$, ΘK, $\Lambda\Xi$, $\Pi\Sigma$, etc, to imitate the circles of longitude or meridians on the Earth's globe, which likewise meet the Equator at right angles. A prime meridian will have to be chosen from these, to begin numbering the longitude, just as on a map of the Earth a prime meridian is chosen, for instance that through the most westerly of the Canary Islands chosen by French geographers in 1634 as the Prime Meridian, or

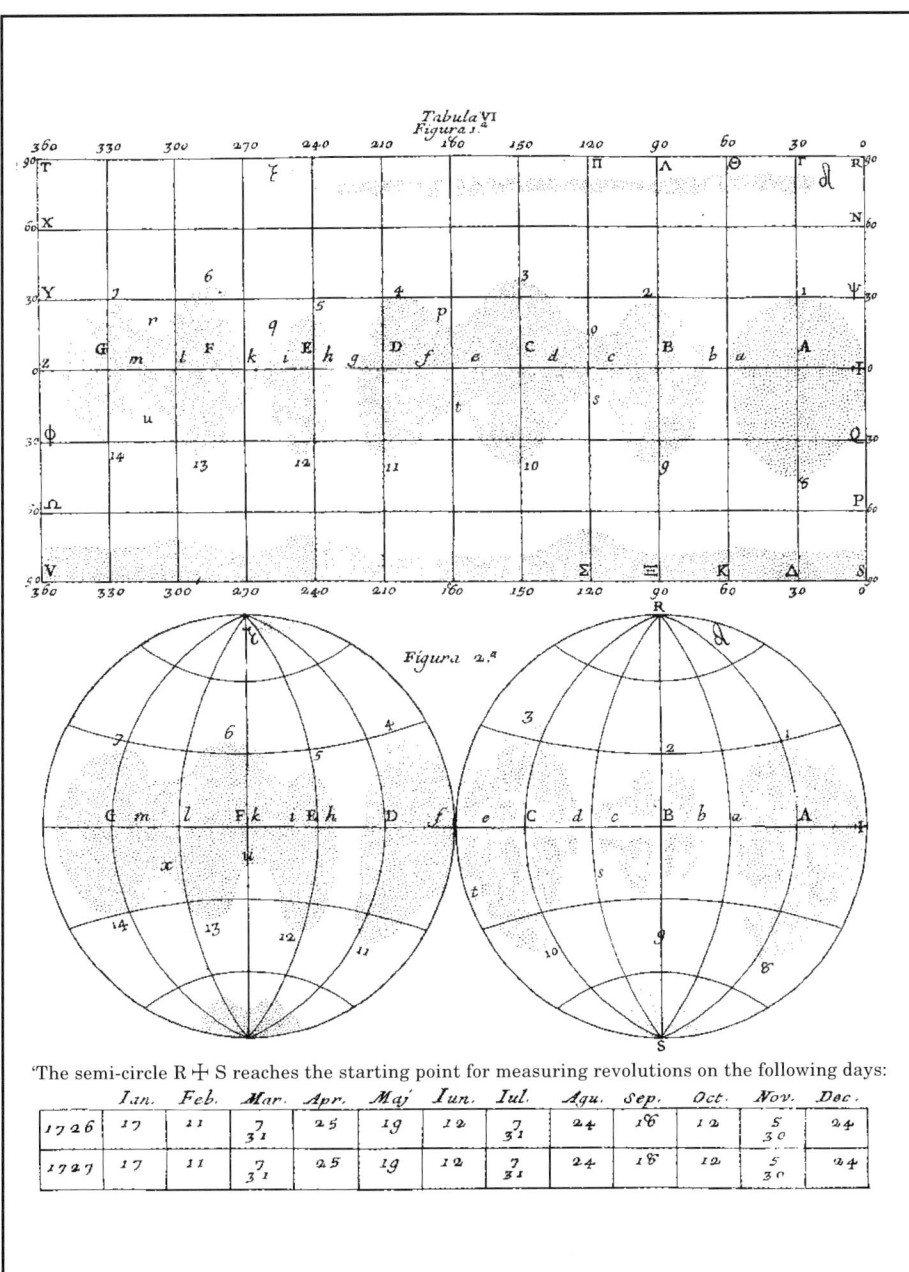

'The semi-circle R ✛ S reaches the starting point for measuring revolutions on the following days:

	Ian.	Feb.	Mar.	Apr.	Maj	Iun.	Iul.	Agu.	Sep.	Oct.	Nov.	Dec.
1726	17	11	7 31	25	19	12	7 31	24	18	12	5 30	24
1727	17	11	7 31	25	19	12	7 31	24	18	12	5 30	24

TABLE VI

that which the Lusitanians have decided upon as the starting point for counting geographical longitude, on the island of the Azores containing the highest mountain, Mount Pico; or that through one of the islands of the Hesperides, nowadays called the Cape Verde islands where the geographers of Holland fix the start of the same numbering on their maps. On the globe of Venus we have selected as the prime circle of longitude the one which runs from the pole along the edge of the round marking A1 (Table II) which was mapped on our first observation of February 9th. I mean along the edge which is nearer to the cusp R than the centre of the disk as shown there, or the one which disappears first below that cusp R, near to which it moves as Venus revolves. That marking A1 is very nearly circular in shape, and about half of it was visible to us on February 9th when we first turned our telescope of 94 palms on the planet Venus. It is not unlike that marking on the Moon called Mare Crisium which is the first to be seen after New Moon, although the one on Venus' hemisphere is much larger, even allowing for the planet's size. We will call it the First, or Royal Sea, for we have been given permission to dignify it with the august name of the most serene and powerful king John V of Lusitania and Algarbia, as we are publishing this small work of ours under the auspices of so great a man. We have decided to call these shaded areas, which are very similar to the lunar ones, 'maria' or seas, and to give them all individual names, derived either from the princes who have prompted the exploration of the New World, or from other famous men in the centuries just past who have undertaken long journeys of discovery in distant parts of the Earth, not forgetting others who have led the victorious navies and land armies of these same European leaders to those places, or from astronomers who have discovered new phenomena in their observations of the planet Venus. Just so did Riccioli adopt the practice of calling lunar markings after famous mathematicians. We have seen that all that area of Venus can conveniently be divided into Seven Seas, which extend on both sides of its Equator about 30 degrees towards the North and South poles, and can be called the tropical zones of Venus. Thus there would remain areas of the planet's globe, from the poles stretching for 60 degrees each side, which would contain two zones, one from 30 to 60 degrees of latitude called the temperate zone, another from 60 to 90 degrees, that is to say to the pole (Table VI), which can be known as the frigid zone. All these zones are shown on the map, crossed from top to bottom by 13 lines, R+S, and those parallel to it, right up to the thirteenth, TZV, which represent circles of longitude. Seven lines cross them, the highest of which, RT, contracts to a point for the North Pole, while the lowest SV does the same for the South Pole, but the middle one, +Z lies on the plane of Venus' equator, and these will be called lines of latitude. It is sufficient to draw these lines every 30 degrees, to avoid confusing the eyes with a multiplicity of detail.

In this rectangular map, where +Z is Venus' equator, +ZYΨ is the northern tropical zone, +ZΦQ is the southern tropical zone, the north temperate zone is ΦYXN, the south temperate zone is QΦΩP, the frigid or north polar zone is NXTR, and the south polar or frigid zone is PΩVS.

VII. Seven seas were particularly observed by us on this planet, covering the middle zone of the globe, ΨYΦQ, which we call the tropical zone. These we have named in order of their curving progression 1, 2, 3, 4, 5, 6 and 7 at their northern extensions, and 8, 9, 10, 11, 12, 13 and 14 where they extend on to the southern hemisphere opposite. In between them we have named four straits (for thus we will indicate the narrow stretches c d, e f, i k and l m), where the markings, which we are calling 'maria' because of their similarity to the lunar ones, join together and interconnect with each other. So the 'maria' will be given names, then the straits, then the northern edges of the straits o, p, q and r, and their southern edges s, t u and x, which we will call promontories since we have decided to borrow these terms from geographers.

The first sea distinguished by the letter A is defined by a perimeter which is almost elliptical, +Ia8. The first measure of its geographical (so to speak) longitude is at the point +, where it is on the meridian selected by us as the planet's prime one R+S and on the equator +Z. But it extends to the 60th degree of longitude, shown at a by another meridian or circle of longitude ΘK. Its northern edge 1 extends to 30 degrees above the Equator, that is to the parallel ΨY approximately, while its southern edge 8 extends further to 45 degrees, about half-way between the Equator and the pole.

The second sea B is bounded by a curved line marked b2c9, the first point of which, b, situated on the Equator, is at longitude 70° counted from the point +, the prime meridian on that same Equator, +Z. The second point 2, the northern bay of that sea, is shown at 100° longitude and 25° latitude north. The fourth point 9 marks the boundary of the southern bay of that sea on the circle of longitude B, that is to say 90° from the prime meridian R+S; while its southern latitude is shown to be on the parallel of 40° S. The third point c is the other limit of its extent in longitude, where sea B connects with the nearby strait c o d s. That boundary is not marked by any variation in light and shade observable on the planet itself, but is an arbitrary division which we have imposed by drawing two lines parallel to the circles of longitude about 10° in front of and behind the northern promontory of this strait o, at the circle of longitude 120°. We will follow the same procedure in the case of the three other straits ef, ik and lm, establishing their longitude as 10° either side of the longitude of the tip of the northern promontory, allotting the letters p, q and r to them individually at the point where the strait is narrower.

We thought it right to follow the same procedure in the division of the seas and straits following, namely the seas C, D, E, F and G. For the three seas B, C and D are joined in such a way that they could be called one sea, separated only from the preceding sea A by a gap, b a, where no shading is evident. There is likewise a clear division, h g, similarly dividing them from the three following seas E, F and G. These latter are also joined, and for this reason could have been considered as a single sea.

But since the sea A has its single northern extension marked 1 and its single southern one marked 8, while the triple tip or northerly extensions of the following marking or sea BCD are marked 2, 3 and 4, and the triple southerly extensions 9, 10 and 11, to show clearly the individual parts of these separate extensions we thought it necessary to count the triple marking BCD as three seas, and likewise to divide the other marking EFG into three, as the figure shows.

Having fixed the number of seas as seven, and assigned limits to them thus in the region ΨΥΦQ of Venus' globe, as well as marking the straits and promontories, to give an accurate description of the planet's Celidography we must proceed to allot individual names to each of the seven seas, to each of the four straits and to each of the eight promontories thus identified in this zone. I will continue with this when I have dealt with the two remaining seas we observed in the polar regions of the planet, one in the north RTXN, and the other in the south PΩVS.

These are adequately depicted in the diagrams already mentioned and explained in Table III, which should be examined again here. The north polar marking is that semi-circular one curving in the shape of a letter C which Table III shows in the diagrams of observations made on July 7th, 10th and 18th. Its curve n o approaches nearest the North Pole at the point indicated by the letter S. However, on this map it has to be distorted into a long streak, for even the pole which should be a point is extended into a line here. We mark the north polar sea with dotted shading in its zone RTXN from the letter δ to ζ. The marking in the south polar region is also shown in Table III similarly for the observation of May 25th, indicated by the letter T at the South Pole as a dotted area. It was not possible to determine its shape as clearly and accurately as the rest, since the distance of Venus from the Earth and its nearness to the horizon where the Ecliptic ascends obliquely hindered precise definition. However, it was seen to consist of four vague and ill-defined small bulges, as I have shown it in the diagram, commending it to other more fortunate observers, who may wish to give a clearer idea of its shape to us and posterity than this rough effort of ours portrays in Table III, depicted as vaguely as it appeared, and distorted in this map because of the conditions imposed by representing a globe in rectangular form.

Therefore seven 'maria' of the middle zone of Venus round its Equator, with their four straits, eight promontories and two polar seas have been counted. We must proceed to assign names to individual ones and indicate their limits or boundaries.

We have followed the examples not only of the old astronomers but also of modern ones in selecting names to indicate individual markings and their more noticeable protrusions and recesses, whether seas, straits or promontories. The ancient astronomers who mapped the constellations on the celestial globe chose their names from outstanding men who excelled others in their knowledge of the stars, or who were the first to demonstrate the dependence of the arts of navigation and agriculture on an understanding of the seasons and the movements of the stars. Thus the memory of Hercules, Chiron and the Argonauts has been preserved on the map of the celestial globe, along with their ships and the memory of other ships too, celebrated in the signs of Pegasus, Taurus, Aquila and Delphinus; Boötes has also been commemorated, Heniochus, Aesculapius and Virgo bearing an ear of corn (Spica), alias Ceres, on account of their cultivation of the land, herds of animals, herbs and trees to provide nourishment and medicine, and the invention of a calendar based on the movements of the Sun, Moon and stars. Modern astronomers likewise have opted to name features recently detected on the planets after famous mathematicians of all ages. Riccioli started the practice with the markings on the Moon that he observed with his telescope, with the approval of the rest of the astronomical fraternity. Alternatively, features are named after princes who promoted these studies by their patronage, as Galileo chose to call Jupiter's satellites the Mediceans, and Cassini copied him by calling Saturn's companions after Louis the Great.

I felt that I should follow the lead given by these latter in honouring the memory of princes, not only those who promoted Astronomy itself, but also those who through this patronage of Astronomy brought great benefit to the nations separated from us by both the Eastern and Western Oceans, for example commemorating the names of the leaders and commanders who set out on expeditions to the Indies under the auspices of that patronage, and not only spread European glory and power everywhere but also (a much more praiseworthy achievement) supported and helped the ministers spreading the divine word, thus procuring eternal salvation for countless nations. I considered that the memory of such great men had even more right to be inscribed in the celestial chronicles than that of the ancient Phoenicians, Egyptians and Greeks, who found a place in the constellations not so much through actual deeds but through legends. I have also reserved an equal amount of space for the two astronomers who have been the first to announce new discoveries concerning Venus, namely Galileo and Cassini, and for the two societies of learned men promoting these studies, the Royal Academy of Sciences which flourishes at Paris[2], and the Bononian Institute for the Promotion of Sciences and the Liberal Arts, founded a little earlier in the Mother City of culture by that great enthusiast for education, Pope Clement XI.

Therefore, while listing the individual markings and naming them in honour of the aforementioned princes, leaders, scholars and learned institutions, I feel it is my right and duty to provide a brief description of the more notable achievements of each, which have earned them already a permanent memorial in history, and now also win them a place in our Celidography.

VIII. The First Sea A, called the Royal Sea of King John

As a loyal servant of his most serene and powerful majesty King John V of Algarbia and Lusitania, in whose prosperous reign our Celidography is being published with the approval and encouragement of this great ruler's own clemency and generosity, providing unstinting patronage to all branches of learning, I could not rightly seek elsewhere for the name of the first Sea that I discovered on the planet. The achievements of this most glorious King are so apparent to the whole world in gathering together the wisdom of all nations to adorn his realm that further praise from me, or explanation of his name taking precedence, is entirely superfluous. But if it were necessary to touch on the achievements of so great a Ruler in striving to secure the salvation of the New World no less than that of the empire acquired by his predecessors, our seas bear more than ample witness to his truly royal religious zeal. For in addition to the expense of the voyage to the Indies, he provided a well-equipped fleet a few years earlier for the defence of all Europe which shared a common danger with Italy in those difficult times when the siege of Corfu by the Turkish tyrant was immediately relieved by the arrival of Lusitanian aid. Therefore let the First Sea receive the auspicious token of his victorious name, equally favourable and propitious to the Christian religion, the fine arts and the sciences, and to be called in our Celidography the Royal Sea of John V.

The boundaries of this First Sea have been carefully defined above, and need no further explanation.

The Second Sea, B, or the Sea of the Infante Henry[3]

The three seas which follow, namely from the second to the fourth, really comprise one continuous marking occupying considerable space from east to west, from 70° to about 216° longitude. But since this marking has

---

[2] *Academie Royal des Sciences, founded by Louis XIV.* **S.B.**
[3] *Henry 'The Navigator'.* **P.F.**

three projections in latitude stretching towards both the poles, with two narrow straits dividing them, the first lying between the Second and Third Seas, the second between the Third and Fourth Seas, similar to the straits which join the seas on Earth, we have divided into three seas, so to speak, the broader extensions of this marking, keeping the name of Straits for the narrower portions in between and the name of Promontories for the extreme tips of these, imitating the nomenclature of geographers. So I will give a description of the seas of this three-part marking in the order of their progression, those of the following one likewise divided into three narrower and wider parts. Then after completing the number of seas on the planet's equator, I will proceed to deal with the individual straits in sequence. This done, the promontories will be considered in turn.

Also in choosing names I have decided a plan whereby the three seas which are joined, the Second to the Fourth, near the First Sea already called Royal, are named after princes of the royal family of Braganza who ruled with prosperity the broad kingdom of the Lusitanian monarchy, and from that band of heroes I have selected those who took the initiative in leading Indian expeditions to both parts of the Ocean, keeping the correct sequence of their ages.

Therefore I will not repeat here the boundaries in longitude and latitude of the Second Sea B, already described. But this sea is called after the Infante Henry to commemorate his expeditions to the Indies, so that he should be considered not only as a supporter and promoter, but in a sense the first instigator of all the voyages undertaken both by the Lusitanians and also by the Spaniards. For although the first discovery of the New World did not occur until almost a century after his death, and the boundaries of the Eastern Ocean had not been reached at that time, nevertheless Henry laid the first firm foundations for undertaking voyages beyond Africa; to such an extent that any successful discoveries and achievements of the following generation owe a debt of gratitude to his foresight and concern, and above all to his zeal in spreading the Christian religion.

Henry was the fifth-born son of the great King John the First. In 1415, in his youth, he accompanied his father on an expedition against the Moors to take by storm the place called Ceuta, and was filled with a burning desire to spread the Faith more widely amongst whatever barbaric peoples were deprived of that light. He obtained from Pope Martin V special privileges for soldiers serving on sacred missions to promote this cause, and devoted all his energies to establishing strongholds that would open up a much safer route for voyages not yet attempted. He gathered together from all sides the most skilful navigators, and even founded colleges of mathematicians not only in Europe but also in Africa beyond the Straits of Gibraltar as far as

Cape St. Vincent[4]. Shortly afterwards he established a commercial law-code, and settled the division of the new acquisitions between the Spaniards and the Lusitanians; then with his reputation well established through all his virtues, and having pioneered a method of achieving brilliant expeditions, he departed this life in 1448, leaving posterity with an enduring sense of loss at his demise.

The Third Sea, C, named after King Emmanuel.[5]

The Third Sea comes next, marked with the letter C on our Celidography map, and to it I give the name of King Emmanuel.

The longitude of this sea extends from the edge of the western strait between this and the Second Sea (about the Strait of Albuquerque I will speak later), that is to say from 130° to 170° longitude. The tip of its northern bay, marked 3, stretches to about 35° above the Equator, while the southern tip opposite stretches to about 40° below it. Both bays reach their widest point at about 160° longitude measured from the Equator of Venus where it meets the first or eastern edge of the First Sea, R+S, the point chosen for the prime meridian of our Celidography.

Regarding the choice of the name King Emmanuel, it is obvious to anyone conversant with the first principles of the history of Lusitania and of the Indies what a debt of gratitude for the prosperity of both Worlds, the Old and the New, is owed to the greatness of intellect, religious zeal, courage in war and foresight of King Emmanuel. The first voyages beyond Africa undertaken in his reign were extended by the truly heroic leaders whom he wisely selected to the remote shores of the Eastern and Western Oceans with amazing success and speed, and are an everlasting memorial to the glory of Lusitania, no less than to the greatness of the Catholic Church. His father was the Infante Dom Ferdinand, son of King Alphonsus V, to whom the name of Africanus was given because of his success. After the death of John II he became the fourteenth King of Lusitania, and won such acclaim for his discoveries and influence in the New World that the glory gained from such bravery seemed just as well deserved as that which he had already won for protecting the Old World against the daring threats of the Turks at the instigation of the Church authorities in Rome. He was on intimate terms with the greatest European kings, and enjoyed particularly strong bonds of friendship with the Asians too, many of whom were his clients and quite a few his tributary dependants. Following the footsteps of the Apostle Thomas,

---

[4] This cannot be the Cape St. Vincent of the extreme southwest of Portugal but an African cape now bearing a different name.

[5] King Emmanuel, 'The Fortunate', reaped the glory of other men's labours. He wished to become king of Spain as well as of Portugal, so he expelled the Jews, the backbone of Portuguese commerce, to please the Spaniards.   **P.F.**

he promoted the spread of the Christian faith to the farthest Indies. When Vasco da Gama set sail from Lisbon in 1497 and, rounding the Cape of Good Hope, continued his journey across the ocean to reach unknown lands, he made a treaty with the King of Melinde and with Zambri of Calicut. He then returned to Europe, but repeated his voyage in 1502 and forced the rulers of Kilwa and Calicut, who had broken the treaty and rebelled from the pact of obedience, to submit again to King Emmanuel. The outstanding achievement of the twenty-six years of his most glorious reign was the extension of his victories as far as the bay of the Ganges. The verdict of historians on his reign is summed up by V.C.Maugin on page 224 of his *'Brief History of Lusitania'*. "He extended his dominion far and wide with the addition of part of those regions which the King of Persia used to rule, and also several Ethiopian states, and these stretched the boundaries of his possessions as far as the Indies, with the result that a considerable part of the realms bordering the River Ganges became subject to him."

His name, therefore, certainly deserves to occupy this place among the princes of the Royal Family who designate the nearby seas, and among the courageous Lusitanian leaders whom we have assigned to the adjoining straits and promontories. For it was particularly under this King of the renowned nation of Lusitania that these Lords of the Sea flourished whom he appointed, as I shall describe after my account of these seven Venusian seas.

The Fourth Sea, D or the Sea of Prince Constantine.[6]

This series of three seas joined together, BCD, is completed by the one we are calling after Prince Constantine, to avoid separating the heroes of the Royal Family famous for their voyages to the Indies.

It shares a common boundary f with the strait which closely precedes it at 190° longitude. Its other limit in longitude is g at about 218°. Its northern extension 4 above Venus' Equator is at about 30° latitude, while its southerly one 11 descends to almost 30° below the Equator, Z, that is the parallel QΦ.

We have named this sea after Prince Constantine, who must be regarded as one of the leading lights of the Royal Family on account of his memorable achievements in the Indian expeditions so outstanding that King Sebastian, by whom he had been sent out in 1558 as one of seven viceroys with full powers, held him up as an example to the other viceroys and put him in charge of them; although the position was offered to him permanently as long as he lived, he declined it. He prosecuted the war with determination against the kings of Janapatai and Manaria, who were persecutors of the Christian faith.

---

[6] *Prince Constantine of Braganza, Viceroy of Goa, in charge of other Viceroys. Offered the post for life, but resigned after the customary three years.* **P.F.**

When King Peguan offered more than 100,000 gold pieces to buy back an idol that had been taken away from him amongst the spoils, he refused vehemently, and, imitating the example of Moses, crushed the effigy into fine dust, to avoid fostering the superstition of that tribe. He emerged from his voyages with the titles of "Fortunate", "Pious", "Just" and "Invincible", with the result that he seemed to communicate the protection of good fortune which he himself was granted by Divine Providence to the ship which was built in the Indies for his return to Europe, since that same ship undertook the long voyage, which in those days was full of danger, ten times and always reached harbour safely, whether in the Indies or in Europe.

The Fifth Sea, E, or the Sea of Columbus.

Along with the Sixth and Seventh, the Fifth Sea constitutes one marking divided likewise into three, with three wider parts divided by two narrower parts or straits, just like the three preceding seas. For that reason I have given them the names of three Italians, Columbus, Vespucci and Galileo. The first two discovered regions unknown to our ancestors on the Earth's globe, and the third revealed phenomena on the planets, that is to say the phases of Venus, crescent, half, gibbous and full. Therefore we have linked together this triumvirate of distinguished explorers of Italian stock, to name the three seas which remain for us to describe around the Equator of Venus, along with their two straits and four promontories which will be dealt with after the description of the seas.

So the Fifth Sea, E, measures its longitude from h at about 228° to i at approximately 250°. The curve of the bay 5 reaches 25°N. latitude , to 35°S. at 12.

The name attributed to this sea commemorates that new Tiphys[7] of the sixteenth century, Columbus, justly called Christopher, a name appropriate to his actions as well as his character. Distinguished alike for his religious devotion, his military skill and his navigational ability, he sought and found a shore beyond the Western Ocean, and was the first to plant there the standard of the Cross of Christ. Brilliant though the story of his voyages and achievements may be, particularly from 1492 when he embarked from Cadiz on September 1st to seek unknown lands, to 1501 when he returned from his last journey to enjoy the rewards of his voyages at the court of Ferdinand, King of Castile it will suffice to mention it briefly here.

The Sixth Sea, F, or the Sea of Vespucci.[8]

The eastern edge of the Sixth Sea, k, is fixed at 270° longitude on

---

[7] *Tiphys - pilot of the ship Argo.* **P.F.**

[8] *Vespucci visited Rio Plata in 1503, mapped from there to Brazil and gave his first name to America.* **P.F.**

Venus' Equator; its western edge at 300° at l. Its northern bay 6 reached its highest point at latitude 35°N, and its lowest at 37° South at 13.

Amerigo Vespucci of Florence, who gave his first name to a hemisphere of the Earth's globe, will justly lend his family name to the Sixth Sea of our Celidography. At the behest of Ferdinand, King of Castile, he undertook two voyages to the West, the first in 1497 and the second in 1499. He also embarked on two expeditions to the Eastern Ocean, on the instructions of King Emmanuel of Lusitania, whom I have mentioned above, the first in 1501 and the second in 1503, which he described himself. As they are the subject of various learned treatises, we are relieved of the responsibility of writing about them in detail.

The Seventh Sea, G, which we name after Galileo.

We measure the start of the Seventh Sea from 320° longitude on Venus' Equator, and its end at about 350°. The most northerly point of its upper bay, 7, is at 25° latitude, and its southern tip we place at about 30° S.

We thought it appropriate to link Galileo's name in our Celidography with those who mapped the Earth's globe, partly because he contributed more, possibly, to geographical accuracy by applying his observations of the eclipses of Jupiter's satellites to an exact determination of terrestrial longitudes[9], than did the voyages of most of the sailors; and also because he was the first to discover, with his telescope, phases similar to the lunar ones in this planet which we are describing.

Names are given to four straits and eight promontories shown in this Equatorial zone of Venus.

The Equatorial zone $\Psi Y \Phi Q$ of our map of Venus' Celidography shows clearly, as well as the Seven Seas so far considered at their broadest parts, where they nearly touch or even exceed the limits of that zone itself, narrower stretches of those same seas like channels or straits, by means of which the seas are joined in both of the three-part markings already described. In the same order, therefore, in which they follow after the First Sea, or the Royal Sea of King John V, which is separate from the rest, we will deal with them in order, fitting names one by one to the full extent of each strait, then also to the promontories which project at the narrowest parts of these straits, preceding the names by an indication of the longitude and latitude of each strait joining the seas on either side.

We have said that the Celidographic longitude of Venus' markings is defined by the same method in our scheme, as we have used to decide the geographical

---

[9] *These acted like a 'clock in the sky'; Harrison's chronometers performed a similar function and allowed longitude to be found, even at sea.*   P.F.

longitudes of places on Earth, that is to say by drawing great circles through the various degrees of the Equator, at right angles to them and meeting at the poles. At the point where the northern edge of a strait is at its shortest distance from the Equator we have drawn one of these lines of longitude at right angles to the Equator, and taken it as the centre of the strait. From this middle point we have counted ten degrees along the Equator in both directions, and marked their lines of longitude. At the first one, that is to say the one nearer to the marking A, we have fixed the eastern limit of that strait, and at the other, further from marking A and nearer the series of following seas BCDEFG, we have fixed the western limit.

So the first strait joining the Second Sea B to the Third Sea C has its centre at 120° longitude, through which the circle of meridian ΠΟΣ cuts the narrowest northern part of the strait at O. If two lines parallel to this line of meridian ΠΟΣ are drawn on the map ten degrees either side, at c and d, the area c o d s is the first Strait, the longitude of which at c will be 110° and at d 130°. The northern latitude of about 11° at O, and the southern latitude of 12° at S, will indicate the two promontories. The first strait we will call after Albuquerque, and its northern promontory o will be Almeide and the southern one, S, Da Cunha.

Alphonso de Albuquerque[10], after whom we choose to name the strait, undertook his first Indian expedition under King Emmanuel in 1503, when he was made commander of the Forces. He struck a treaty with the king of Cochin, founded a church and a stronghold there, and ordered three ships to defend that same king against the power of Zamora. In the second expedition, in 1506, he had as his ally Tristan da Cunha, who set out with him to establish defences in the island of Zocotora, and to protect the Christians of Abyssinia against the incursions of the Moors. Indeed, Tristan was in charge of the Indian fleet, and Albuquerque of the troops. After they captured the stronghold of Zocotora from the Moors and furnished it with new defences and a strong garrison, he went to Arabia and established laws of commerce at Calaya. He made the King of Ormuz a tributary to Lusitania. Finally, while a Prefect in India, he acquired three well-populated cities to be under Lusitanian rule, Ormuz, Goa and Mallacca, so that his fame resounded far and wide throughout Egypt, Persia and the Indies. He cultivated the friendship of their most influential kings by the sending and receiving of embassies, which greatly benefited the Christian religion. At last, loaded with honours, he departed this life at Goa on December 26, 1515.

---

[10] *The most famous of all who served in India, he led a successful expedition to the East Indies and a less successful one to Aden; he was popular with the Hindu rulers because of his fair treatment.* **P.F.**

The northern promontory of this strait, marked with the letter o, we name after Francisco de Almeida[11]. He was the first to be honoured with the title of Viceroy in the Indian acquisitions in 1505. While sailing along the African coast with his fleet, he brought the King of Kilwa back into line, and disciplined the rebellious Zamora with severe punishment, but received the suppliant King Onor as a client. He set up several strongholds to safeguard navigation at Mombaza, Archeliva and Cannanor. Indeed, he deserved a more propitious end to his life than that which he experienced, since while returning from the Indian expedition he was killed by the Cafri at the Cape of Good Hope.

At the southern promontory of this strait we have commemorated Nuno da Cunha[12], who must be regarded as the second Albuquerque. He set sail from Europe with his father Tristan in 1506, besieged Mombaza on the African coast and made its king pay tribute to Lusitania. Resuming his navigation, he set out to undertake the duties of viceroy in 1528. He received the submission of the King of Ormuz but refused to take bribes. He punished the treachery of the King of Cambaya with death, and from his successor received not only peace, but also power to found the city of Diu, since he had previously established the royal fort of Barace or Bazaim in the kingdom of Guzaratis. He made the celebrated Antonio de Sylveira ruler of Diensi (Diu), the city he had founded; you might call de Sylveira the Horatius Cocles[13] of Lusitania, for he fiercely defended the city single-handed at the first siege, when it was cut off by land and sea from any hope of relief from its allies: by land by the Cambayan forces, and by sea by the fleet of the Turkish tyrant Solyman, numbering seventy triremes and seven thousand fighting men. Cunha too departed this life on his journey back from the Indian expedition, though worthy of a more illustrious tomb.

The other strait follows between the Third Sea, C, and the Fourth, D. The middle point of this strait is on the 180° line of longitude, while it passes through its narrowest point at the north, p, at about 12° latitude. The narrowest point south, however, is at 18° latitude below the Equator. Therefore, the area of this strait, e p f t, extends from 170° to 190° longitude.

We have borrowed Vasco da Gama's name for this strait. The northern promontory, p, has taken its name from Duartes Pacheco, while the southern one, t, we have called after John de Castro.

---

[11] *First viceroy established by Emmanuel on the Malabar coast.* **P.F.**

[12] *Son of Tristan, the Great Navigator, falsely accused of cruelty and dismissed.* **P.F.**

[13] *Famous Roman soldier who defended the Tiber bridge single-handed against the Etruscans.* **S.B.**

Vasco da Gama was sent by King Emmanuel in 1497 to sail round the Cape of Good Hope and explore the East Indies. He encountered countless difficulties on his journey, and met all dangers with a determined spirit. He made a pact of friendship with the King of Melinda, and was honourably received at Calicut. He also struck a treaty with Zamora before returning to Europe. He set out again in 1502, dignified with the title of Commander of the Fleet. From the King of Kilwa he exacted the submission that had been denied to the Lusitanian authorities. He tamed the rebellious Zamora and sailed victorious into the port of Lisbon with thirteen ships after slaying the enemy commander and exacting tribute from the King of Kilwa. He then undertook a third expedition, in 1524, honoured by the title of viceroy, but died a few months later. His descendants inherited the legacy of his successes, for they were awarded the perpetual title of Commanders of the Indian Ocean by King Emmanuel, and these honours were increased by succeeding generations, who won the right to enjoy the privileges of the Companions of Vidigueyra, and a few years later those of the Lords of Nicaea.

Duartes Pacheco embarked along with Alphonso Albuquerque on the voyage to the Indies in 1503. The latter left him at Cochin with three ships and a small band of Lusitanians, numbering only 60, to guard the King of Cochin with whom he had a treaty, against 80,000 men led by Zamora. He displayed such courage and wisdom in the disposition of his troops, consisting only of his own small number and 300 Malabarians, that he confronted the enemy in a narrow pass and devastated their huge army[14], ensuring the safety of the allied king until reinforcements could be sent by the Lusitanians. On his return to Lusitania he was honoured so greatly by King Emmanuel that he was later accused of conspiracy by those that were jealous of him, and held in prison, where he maintained his innocence with Christian constancy, and endured an entirely undeserved conclusion to his life.

John de Castro[15], the fourth viceroy in India, had his baptism in military service in Africa, the customary arena for the Lusitanian heroes. He set out for the Indies in 1538, content with the command of one ship, and refused the governorship of Ormuz, feeling that he had not earned it. Seven years later he was sent to govern the whole province, which the King of Cambaya was invading, having conscripted a strong army of select troops loyal to him, reinforced by auxiliaries from the Turks, Abyssinians and all

---

[14] *This victory secured his fame and more rulers became interested in trade with Portugal.* **P.F.**

[15] *Castro served with the Emperor Charles V, distinguished himself in an expedition to Tunis, later served in the Indian seas and supervised the evacuation of unwanted outposts in Morocco. Viceroy of India in 1545, he was victorious at the siege of Diu and cleaned up the civil administration.* **P.F.**

the disruptive elements bent on the overthrow of the state of Diu and the expulsion of the Lusitanians from those shores. When the illustrious Duke of Mascaregnas had bravely endured an eight-month siege, John de Castro arrived just in time with a fleet of ninety ships and 2,500 troops, and immediately attacked the enemy camp and after enticing their army onto the open plain, annihilated it. The pressures on the Royal Treasury, drained by so many wars, prevented the occupation of the enemy's fortified posts, for the expense of a siege was unbearable. The Senate of Goa and the ladies of the Court vied with each other to offer the Viceroy their valuables to swell the Treasury's resources. But John, with equal magnanimity, sent everything back once he had managed to pay the military wages and furnish the resources for continuing the war from the enemy spoils (obtained from a richly-laden Cambayan ship captured on its return from Persian harbours). So he was accorded a triumphal reception by the people of Goa as a leader of outstanding courage, yet showed great restraint in that while not belittling the praises his men bestowed on him as due reward for his bravery and patriotism, he went straight back to do battle with the rebellious forces who were relying upon the support and influence of Hildacan. But against both armies this spirited commander moved with such speed and put up such a bold and successful fight that the fame of his victory spread far and wide. The King of Narsinga was induced to seek a treaty of friendship with him. He reckoned that he would have to renew hostilities with the King of Cambaya, who was advancing on the city of Diu with an army of 150,000, although his own troops numbered only 3,000. But the barbarians refused to fight, and sounded the retreat. The arrival of John de Castro was sufficient to make them flee in terror.

Castro's military prowess was allied to an overwhelming desire to promote the spread of Christianity, in that he followed the example of St. Francis Xavier, a missionary in the Indies, by cultivating every kind of virtue in his own life. The mission to Ceylon traced its origin to his administration, as did several other undertakings. Finally he died in the embrace of this same Xavier, with great piety, on June 6th, 1640. Although a Viceroy of India, nothing was found in his domestic treasury except some instruments of self-mortification, and three silver coins. He had distributed among the poor his official salary and all his wealth which accumulated because of the frugal life-style of this most abstemious of leaders.

The Third Strait lies between the Fifth Sea E, called after Columbus, and the Sixth F, called after Vespucci. Its centre is taken from the line of longitude through the tip of the northern promontory nearest the Equator, whose Celidographic longitude is about 260°, and latitude 10°N. The area of that Sea, i q k u, has its eastern limit at 250° longitude and its western limit at 270°. Its southern promontory's latitude is similar at eight or ten degrees.

I have named the Strait after Cortez, the northern promontory after Pizarro and the southern one after Chabral.

Ferdinand Cortez was the founder of the Spanish possessions on the American continent. He originated from Fort Metallina, commonly known as Medelino in the Spanish province of Extremadura, and in command of a fleet of ten ships he landed at the port afterwards called the Harbour of the True Cross, in 1519. He won control of the Kingdom of Mexico for Charles V, after first striking a treaty with Montezuma, who held sway in that area, but soon afterwards taking him prisoner, since the barbarian had broken the pact of friendship by murdering some of his men. He sent King Charles a huge amount of gold from the spoils. Yet he swelled the treasury even more from his subsequent victories. He acquired 19 more ships, and was able to take over the army sent against him by Didaco Velasquez, and dispatch ships and soldiers wherever he wished to gain control of the whole region of Mexico, including the capital itself, where he made his victorious entrance on August 13th, 1521. He abolished their barbaric superstitions by destroying their enormous effigies and temples, where human sacrifices took place. On his return to Spain he was honoured by Charles V with the title of Marquis de Valle, whose income today is estimated at sixty thousand, and also the rank of supreme military commander, or Captain General of New Spain, as it is called. Other honours were heaped upon him, and he continued his voyages to America with the discovery of California. Back in Europe again, he wished to follow in Caesar's footsteps with an expedition to Africa, but met his death at Hispali on December 2nd, 1545 at the age of 62 as he reaped the rewards of his great achievements. His outstanding deeds are mentioned not only by the Spanish historians but also those of the Church, such as Spondanus. Above all, J. Francis Gemelli has published a summary in his "Survey of the Globe", Part VI, book 3, chapter 4, where he has drawn up a brief but very readable account of the discovery and conquest of New Spain.

Chapter V of the same work deals with the history of the conquest of Peru by Francis Pizarro under the auspices of the Emperor Charles V, from whom he also obtained command over Peru and a Supreme Prefecture of New Castile, or the position of a Captain General. But although the earlier actions of Francis were not marred by a single crime, later he was motivated by insatiable avarice to act treacherously towards the unfortunate King Atabaliva, and then rebelled even against Caesar himself, in whose name he was beheaded in 1548. I was extremely dubious about retaining the memory of such an evil man amongst the famous leaders of expeditions to the Indies. However, to avoid a hiatus in the history attached to our Celidography if mention of the conquest of Peru were omitted, I have made a concession to the need for a complete history which I would have denied to the wicked polluter of that glory which he had won for himself at the outset of his remarkable voyages of discovery.

The reputation of history is upheld, however, by the name of Pedro Alvares Cabral[16], or Capralis, at the South Promontory of this strait, who completed Lusitanian conquests, at least on the coast of America, with the addition of Brazil to their empire. He was the first to discover this country when Divine Providence drove his ship there after a violent storm forced it away from the African coast, in order to bring humanity and the light of Christianity to these savage peoples. This was furthered conscientiously by the most reverent Kings Emmanuel and John of Lusitania when they founded cathedral churches in such far-flung parts of the globe, and dispatched sacred ministers there selected by the Pope, through whose work and teaching these primitive tribes became civilized. But Cabral augmented the glory won in South America with fresh expeditions in Africa and Asia, with the result that he escorted back to Europe ambassadors sent to him from Cochin and Cannanor in Malabria.

So far, therefore we have listed six Seas, three Straits and six Promontories consecrated to the memory either of great rulers or famous leaders who were the first to open up a considerable part of the globe to navigation.

The Seventh Sea, with its adjoining Strait and the two Promontories projecting on to the Strait, I have reserved for distinguished astronomers, whose observations have made the greatest contributions to those same terrestrial voyages which daily are adding to our knowledge of Geography, and whose new discoveries have opened up the Solar System to ourselves and to posterity.

The prince of them all is Galileo Galilei, the nobleman of Florence, who increased our knowledge of Mathematics and Physics with so many new discoveries about the laws of motion, the resistance of solids floating in liquid, the oscillations of the pendulum and similar things. By the invention of the telescope he brought the heavenly bodies closer to Earth, discovering sunspots and the satellites of Jupiter, whose eclipses and occultations have helped us to map the Earth's globe more accurately and thus travel more safely. Finally, he was the first to claim, or at least to demonstrate, that Venus shows phases similar to those of the Moon. So I thought that the seventh Sea should bear his name; its longitude on its eastern edge is 320° at the Equator, and at the western edge about 350°.

---

[16] *Discovered Brazil by accident. Retraced the route of Vasco da Gama to Calicut.* **P.F.**

Its northerly extension reaches 27° latitude above the Equator, and its southern extremity dips 30° below it.

The strait to which Cassini gives his name joins this seventh Sea of Galileo and the sixth Sea of Vespucci. The mid-point of this strait is fixed at 310° longitude, which makes its eastern boundary 300° and its western one 320°.

It is right to commemorate J.D.Cassini here, since the first indication of one or two markings on Venus comes from a letter of his, not actually published by him, but circulated in his lifetime both in the learned "Ephemerides", and in the "Celestial Sphere" of Ozanam, as will be mentioned below on several occasions. It was also through the hard work of Cassini himself that the tables of the movements of Jupiter's satellites were refined to such a pitch of accuracy that they surpassed all other methods of ensuring safe navigation and thorough knowledge of Geography[17]. Therefore on more than one count he has earned the right to a place in our Celidography.

The discoveries by these illustrious mathematicians have prompted the foundation by great rulers of several societies for the advancement of Science. Two of these, however, are more closely relevant by dint of their original connections with Galileo and Cassini, and for that reason we have thought it appropriate to give their names and associate their merits to the two promontories bordering the Strait of Cassini and the Sea of Galileo. The first of these two societies is the Royal Academy of Sciences which Louis the Great founded in Paris. This great patron of learning enticed Cassini there to blend his Italian genius with the cream of French intellectuals. The other society was set up at Bononia by Pope Clement XI, a great enthusiast for the fine arts, whose generous support has encouraged the best literary efforts of the most outstanding noblemen, senators and other men of rank in that state. It is called the Institute of Sciences and Fine Arts, and justifies its country's ancient title, the Mother of Culture. Astronomy is particularly well provided for in both these institutions, for Paris boasts an observatory magnificently appointed and furnished with all manner of instruments through the generosity of Louis the Great, who chose as its directors the most famous astronomers of the time, Cassini and de la Hire, to be succeeded by others clearly worthy of their mantle, supported by their families and descendants[18]; while at Bononia, in addition to the fact that Cassini worked out the greatest helioptic Meridian line in the spacious church of St. Petronius, an Observatory has recently been set up under the management of the renowned Eustace Manfred at his house, and it is second to none of those which

---

[17] *Longitudes of overseas places were indeed successfully found by this method, but unfortunately it was useless to mount a telescope on the deck of a heaving ship at sea.* **P.F.**

[18] *There were 3 more generations of Cassinis at Paris observatory.* **P.F.**

the generosity of great princes has provided for the advancement of Astronomy. Galileo indeed inflamed the Italian academies with enthusiasm for celestial observations, and at the same time was torchbearer for the whole world in pioneering the use of the telescope, contributing in addition many other measures for developing these studies. Thus a method was born in Italy and developed first within and then beyond its boundaries of promoting astronomy with more success in one century than in all preceding ages. Therefore the naming of these promontories in our Celidography after the Royal Academy of Sciences and the Bononian Institute, along with the Strait of Cassini and the nearby Sea of Galileo, has provided a suitable conclusion to the nomenclature of the series of markings which clothe the equatorial region of Venus to around 30° latitude north and south.

IX. In both the polar regions of the planet two seas remain which need to be given names, and whose boundaries must be defined. At the north is seen the polar marking shown earlier in Table III in the shape that it appeared from the Earth in our observations of July 1727, marked with the letters n o p r z round the Pole. It is distorted, however, on this map, which shows as parallel lines circles which meet at the Pole, so that its latitude and longitude correspond to the degrees correctly shown on the globe, with its name added, the Northern Sea or the Sea of Marco Polo.

The eastern edge of this semicircular sea is at about 35° longitude, and is distant from the North Pole 20 to 26 degrees; therefore its latitude measured from Venus' equator extends for 6 degrees from 64° to 70° N. The other edge, which is nearer to the North Pole, is at 265° longitude, while its latitude covers about 8 degrees extending from 5° to 13° from the Pole, which is equivalent to 77° and 85° from the Equator.

It seemed appropriate to name this sea, which is the nearest to the North Pole of Venus, after the Venetian nobleman Marco Polo, who was the first to undertake journeys to the east coast of Asia in the direction of China, and paved the way for those other expeditions that we mentioned a little earlier.

For after a voyage to Constantinople in 1269, when it was ruled from the West by Baldinus, on his return to his homeland he was filled with a burning desire to undertake a journey to the Eastern lands. So he set off to the court of the great Chamus of Tartary where he was received so honourably that he was even sent by that same prince to the Pope to obtain priests to instruct the peoples under his rule in the Christian faith. He was the first to supply us with any knowledge of the Far East and the islands of the Indian Ocean. Scanty though it was, it was enough for the intellects of the time to grasp, in that they were not yet stimulated by a deeper awareness brought about by communication with the Indies, and adequate to motivate the generations immediately succeeding them to penetrate those regions which Marco Polo had attempted to explore on his land journey.

In the southern hemisphere of the planet, however, the polar region is occupied by a marking about 20° from the pole which is more or less straight except that I have noticed three extensions a little further from the pole. One is at longitude 120° and extends 35° from the pole towards the Equator; another at 240° longitude and stretching almost 30° from the pole; lastly the third one is at longitude 310° and about 28° from the pole. However, I cannot claim to have seen these extensions as clearly defined as the rest, since they became visible to us at the end of May 1726 when Venus was a little further from the Earth, and it was more difficult to observe in the morning twilight in a part of the Ecliptic which ascends less steeply, so that it did not reach the height above the horizon that it did in the evening observations of February and March that same year. This resulted in the marking or sea appearing like a little cloud with a wispy edge, which gradually faded out between 15 or 20 degrees from the pole at its narrower part, and 25 to 30 degrees where it appeared wider. This we have shown in Table III, symbol T of the observation of May 25th. For this reason we propose that the shape of this southern polar sea should be subject to correction in future observations by others or ourselves. I name this the Sea of Magellan, because it has a similar location to the south polar zone on Earth which Ferdinand Magellan first discovered, and it is called after the Magellan Strait which he was the first to traverse, but which is still little known, in that it is not on the regular shipping routes of today.

If our celidographic map of the markings of Venus deserves consideration, therefore, the point should be made that it is more suitable for showing those in the middle or tropical zone of the planet, 30 degrees to either side of the equator, than the others in the polar regions which would be better shown on a planisphere, but best of all on a globe which presents the solid shape of the planet.

But even in the case of markings, or seas, straits and promontories which occupy the aforementioned zone nearest the Equator, the precision of our observations is not sufficient to define degrees of latitude and longitude exactly, although these markings particularly were presented to our view when the planet was close and they were well placed on the lit hemisphere. We consider that we have done justice to our own efforts and met the requirements of other observers if we admit a margin of error of no more than four or five degrees in the measurements that we have given.

X. After marking the positions of the seven principal seas on the rectangular map, it remains to show them on the planisphere too, since on it a more accurate portrayal is possible even of the polar markings or seas.

Among the several methods employed by geographers to show the Earth's globe by means of two planispheres, we have selected the one which the Royal Geographer, Nicolas de Fer, preferred to all others in that the true interrelationship of the various parts of the globe could be shown with the least distortion. It was invented by this renowned authority, and is a representation of the Earth's globe where curved lines meet at each pole, but not however obeying the laws of perspective by imagining the eye of the observer to be at a point on the Equator (as other astronomers and geographers had done), but instead modifying the theory of global sections to achieve equal distances on the globe between equal differences of latitude and longitude, as can be shown on the planispheres which he published from 1700 onwards.

We followed this method in drawing the other circular map, which shows Venus' globe divided into two hemispheres, with its markings or seas having the same size, with their individual curves and boundaries corresponding to the same latitudes and longitudes that we have already shown on the rectangular map previously described and explained. The letters and numbers also correspond on both maps, referring to the same markings and their individual parts divided into seas, straits and promontories.

In this representation by means of two hemispheres, however, we had to separate Sea C, which is the last one in the first hemisphere, from the Fourth Sea, which is in the second hemisphere although it is joined to the Third Sea. It is not possible to join two-dimensional circular figures at e and f. But the reader will realize that they continue from one hemisphere to another, as seen on a solid globe.

XI. On a solid globe, certainly, there will be no distortion to cause the model to differ from the original globe of Venus, and both its markings and its phases will be faithfully represented. Therefore there will be complete agreement in all of these if a solid globe is constructed which will revolve round its poles, and on which the circle of the Equator, half-way between them, should first be marked and divided into 360 degrees distributed equally among 12 signs, as is customary. At the beginning of each sign, circles of longitude at right angles to the Equator should be drawn and meeting at the poles in the order indicated on the map, and every 30-degrees from the Equator to the poles the circles of latitude should be drawn parallel to the Equator. Regarding the above mentioned circles of longitude, the positions must be worked out of objects placed within any of these 30-degree divisions; and in a similar way it is sufficient to show the start of each sign on the Equator of Venus at 30-degree intervals, so the intermediate positions can be gauged adequately. Having done this, we can proceed to indicate on the globe the seas or markings of Venus with their curves and bends.

Choose any line of longitude, and where it crosses the Equator the sign ♈ should be marked, as a starting place from which the rest are deduced. From the list given just previously with the names of the markings and their curves, the latitudes and longitudes of the edges of each can be found out. Therefore the latitude and longitude given in that list can be transferred to the globe with the lines of latitude and longitude drawn on it. Then the globe will be a three-dimensional model of the desired planet with its celidography. If an axis of revolution is fixed through the poles and set at an angle of 15 degrees to the plane of the Ecliptic and on a line parallel to the radius of its orbit which cuts the 20th degrees of Leo and Aquarius, and the globe is exposed to the light of a candle, lit at the centre of the planisphere, (marked A in Figure 1, Table V), to represent the Sun shining at the centre of the planet's orbit, the illumination of the globe will be similar to that of the real Venus on that day. So also the markings carried on that day to the illuminated hemisphere by the planet's rotation will appear on the fictitious globe just as on Venus itself, if the markings on the fictitious globe are brought to a similar position relative to the terminator as those observed on the real globe.

This imitation of the real phases of Venus, however, will proceed with greater success and clarity if a crystal lens is placed between the candle and the globe inscribed with the markings to collect the light of the candle and direct it towards the little globe. For then the terminator will be more clearly defined, as more rays will be collected by the dioptric lens, and the dark hemisphere will be almost invisible in contrast to the bright one nearby (as long as there are no nearby objects reflecting secondary light on to it); and the shape of the planet will appear half or crescent on the little globe just as in the sky, if the spectators are placed where the lines from their eyes to the globe and to the candle form a similar triangle to that which lines from Earth to Venus and to the Sun form.

I will refrain here from giving directions for drawing on a flat chart sections of the globe which can be cut out and stuck on to cover a whole globe, exactly representing the markings in sequence, in addition to the other methods of showing the markings we observed on Venus by maps, planispheres and globes. I have good reason for keeping silent about these directions, for they are no different from the familiar rules already taught by mathematicians for constructing both celestial and terrestrial globes.

CHAPTER V

CONCERNING THE PLANET'S ROTATION ROUND ITS OWN AXIS, OR PERIEILESIS (Περιειλησις) IN A PERIOD OF TWENTY-FOUR DAYS AND EIGHT HOURS.

Summary of the Chapter

I. *The difficulty of constructing telescopes of 100 palms and directing them to the Heavens before the years 1680-1700 prevented astronomers from observing markings clearly on the planet Venus even after the detection of one or two, which Cassini claimed to have observed on the globe in 1666 and 1667, since such large instruments are needed to distinguish them accurately.*

II. *After these years, Cassini did not publish anything at all for the whole 36 years that he survived about the markings on Venus or the true measure of its revolution about its own axis.*

III. *This revolution is not completed in 23 hours, as others thought, but requires 24 whole days and some hours.*

IV. *The measure of the time required for one revolution is shown from our observations, especially those of February 26, 1726.*

V. *The sequence of observations on that day.*

VI. *The conclusion necessary from this, that Venus completes one revolution round its own axis not in 23 hours but in 24 days.*

VII. *After establishing satisfactorily the number of days needed for one revolution, we proceed to enquire by means of other observations how many hours must also be added to that whole number of days to achieve greater accuracy.*

VIII. *From several observations in 1726 and 1727 it is concluded that one complete revolution of Venus round its own axis requires 24 whole days and 7 or 8 hours in addition.*

IX. *A more accurate measure of the hours can be made after another 8-year period by undertaking more exact observations.*

X. *Meanwhile the period of $24^{1}/_{3}$ days, or 24 days 8 hours, is taken and a table is drawn up to show at any given time the positions of the markings we observed on the disk of Venus, and also the longitude of its central meridian on a given day.*

I. The description of markings or seas like the lunar 'maria' which appear on the globe of Venus which I gave in the preceding chapter from observations of 1726, had already been attempted by famous astronomers, especially Cassini about 1666, but subsequently interrupted, and even abandoned completely, because of defects in the instruments, in my opinion. At that time Guiseppe Campani and Eustace Divini[1] had not yet built telescopes of the size that we said was necessary to see the markings clearly. For though they were the most skilled in optics, they had difficulty in extending the focal length of the object-glasses to 50 or 60 palms; not only was it a problem moulding glass to the required shape, but also the task was no less formidable of constructing such a long tube which did not bend under its own weight nor curve in the middle when the eyepiece was so far from the objective, particularly when the telescope was lifted above the horizontal to focus on the stars. A clear proof of this second difficulty is given by two devices tried out at Rome around that time and in the years just following. The first was that of R.P. Gottignez, who taught mathematics at the Roman College with great distinction, and the second that of Guiseppe Campani, whose device we saw in 1684 when he set it up in the Pamphilian Gardens outside the Janiculum Gate, which allowed a tube of about 70 palms to be used successfully for the first time for the observation of lunar markings. We include both devices as shown at the time on brass engravings, so that the difficulty of handling such a machine is obvious to everyone, particularly when it is extended to 100 or 200 palms. But Christian Huygens solved this second problem in 1680 when he invented a method of bringing the lenses of a telescope to the same focus without the need for a tube by the simple device of a thin silken thread. With the help of this, the object-glass and eyepiece are held in the same straight line with the object being viewed at whatever elevation it may be, as the axis goes through them. However, this invention was hardly put to much practical use right up to the year 1700, at least not with telescopes of 100 palms (we know of hardly anyone except Huygens and Campani who tried their hand with such instruments at that time), and we have seen that these are necessary to show the markings clearly on the planet Venus. (See Tables VII and VIII.)

II. Cassini had in fact been the motivator when Campani, supported by the generosity of Louis the Great, succeeded in building those huge telescopes whose object-glasses had focal lengths of 100, 120, 150 and 200 palms. Campani had sent them from Rome to the Royal Observatory, where Cassini received them about 1682, and they were later returned to Rome shortly before the death of Queen Christine of Swabia in 1689. Although he discovered another

---

[1] *Eustachio Divini at Bologna and Guiseppe Campani at Rome were the two foremost lens and telescope makers in Italy at the time.*   **P.F.**

four satellites of Saturn besides that of Huygens with them, he could not accommodate them himself when the Huygenian method with its later additions was perfected around 1712. At that time, we brought these additions to the learned Fellows of the Royal Academy, and they were received with such approval by them (which is typical of their generosity) that on their orders they were recorded that same year in the Academy's historical records. But in that same year, while staying in Paris, I performed the last rites for the dying Cassini, whose great contribution to astronomy had been second to none, and with whom I had enjoyed a prolific and long-standing correspondence since before the end of the 17th century, when owing to the pressure of work at Rome I handed over my astronomical research to him. Yet this outstanding man suffered complete loss of eyesight towards the end of his eighty years of life, which was a great blow to the sciences which he had brought to such a peak of achievement, with the result that he could not use these telescopes on Venus at the time when the method of using the Huygenian devices had been perfected and publicized. Certainly I do not remember hearing him mention anything about the markings on Venus and its spinning or rotation around its own axis, or writing anything new in his letters except the rough beginnings of observations said to have been made in 1666 and 1667. Nor were these published by Cassini himself, but only made known in his private letter to Master Petit, and quoted in the "Journal des Savans' in 1667, in the edition published at Amsterdam in 1679 (the only one I have seen), Volume 2, page 257. Then they were mentioned again by the renowned Ozanam on page 80 of his *"Celestial Sphere"*. At the end of this chapter I will append the letter sent to me by the Very Reverend P. Melchior de Briga, a Jesuit priest, soon after I had met him in Florence in 1726, renewing our old friendship, and I had told him of my observations of Venus made that same year. He said he was preparing a report on his many observations of this planet, and together we reviewed the difficulties experienced by astronomers in observing markings of this kind, and talked of the successful attempt to overcome these difficulties which I had made that year. He was the first to make known to me that extract from Cassini's letter to Master Petit, and the mention of his observations of the Venusian markings in the book of Ozanam translated from the *"Journal des Savans"*, when he took me to the library nearby belonging to Marquis Riccardi where he remembered reading it. In his subsequent correspondence with me he wanted to record what interchange of views we had on that subject. Furthermore he sought out the 1679 edition of the Journal, where a fuller account was given of the matter, reconsidered the statements given there in the light of his own carefully-pondered conclusions, and then gave me permission to publish it just at the right time to make this celestial history of my discoveries more complete.

Thus I have studied Cassini's letter of 1667 sent privately to a friend who shared his interests, and bearing in mind the difficulty of seeing the Venusian markings after sunrise even with 100 palm telescopes, when the letter indicates they were seen after sunrise and we know for certain that the telescopes in use were much shorter, and considering Cassini's own silence for the whole 36 years that he survived from 1667 to 1712 on the subject of these markings, I am inclined to believe that he had reached no firm conclusions from the experiments he conducted.

III. But as regards the planet's rotation round its own axis in 23 hours, which the *Journal des Savans* and the writings of Ozanam claim to be true, and which Halley and other astronomers follow when giving its rotation period, Cassini, who was very careful indeed about what he said, came to no firm conclusion in the writings he made public, so far as I know. For he would not be able to decide from any orderly change or progression of the markings across the disk of Venus whether a full revolution was completed within 23 hours or rather within 24 days, unless such an opportunity of observing the planet at a very close approach to the Earth was presented that for three whole hours before the rising or after the setting of the Sun it was well seen above the horizon. This was proved from our observations of 1726, when we established that a whole revolution of Venus round its own axis takes not 23 hours, but 24 complete days.

IV. I said in Chapter II, page 32, that from February 9th for several days following, we observed Venus around the time of evening twilight, and saw not only the markings but also their progress across the disk, which we measured as about 15° per day from west to east.

Therefore, two possible conclusions can be drawn from this. The first one is that within 24 hours the planet has completed one rotation and has begun a second by turning a further 15°, that is to say a twenty-fourth part of the full circle. The second is that within 24 hours the markings have only turned 15°, a twenty-fourth part of the whole circumference, and thus will complete a single revolution in 24 days, just as the movement of sunspots shows that the Sun turns on its own axis once in 28 days.

Which conclusion is to be preferred could not be decided without the help of another observation, which I conducted after first taking the following argument into consideration. For if the globe of Venus, I said to myself, completes a full revolution within 23 hours, it must necessarily turn through a quarter of a circle in 5 hours 45 minutes, and an eighth of a circle in not quite three hours, namely 2 hours 53 minutes 30 seconds. Now indeed, when I was observing in February 1726 Venus was visible after sunset above the horizon for more than $3^1/_2$ hours, so if I noted the positions of the markings at sunset, a gap of $3^1/_2$ hours could be left before making fresh observations to find out whether the markings which had appeared in the centre at sunset had completed an eight of a circle in the meantime, and moved from the centre to the nearest cusp by at least a quarter of the diameter. If a change of a quarter of the diameter could be seen, the conclusion should be drawn that the eighth part of circle covered in this time indicated a full revolution in 23 hours. But if there were only a small change in 3 hours, that was hardly noticeable, it should be obvious that the markings had only traversed a few degrees with the planet's rotation and would need a full day to complete the daily progress of 15 degrees, and a whole revolution would require 24 of those daily amounts and therefore take 24 days.

We enjoyed one evening and night that was particularly suitable for this observation, on February 26th. The sequence of observations on that day I will describe from the Journal of Celestial Observations which I have completed almost daily ever since I traced the Meridian Line from the Baths of Diocletian to the shrine of Maria of the Angels on the orders of Pope Clement XI for the purposes of the Calendar. Here therefore is that sequence of observations.

V. "At Rome on the third day of the holidays, February 26th, 1726. The sky was very clear and the air steady without any breath of wind. In the Palazzo Barberi on the Quirinal Hill, on the wooden bridge which joins the flat area of the gardens to the palace of His Eminence Cardinal Francisco Barberi, we set up the support of an object-glass of 88 palms made by Guiseppe Campani, and then we attached to it a silken thread, extended according to the Huygenian method so that the eyepiece was the right distance from the object-glass, and the axes of both kept in the same straight line from the lower level of the Palazzo Barberi extending up to the fort-like structure with the moat. I began to observe Venus from sunset onwards for almost a whole hour, namely to 5.45 pm., with several others standing by and taking turns to confirm my observations of the markings which I describe.

"Venus appeared as a crescent, as I show in Chapter II, Table II, February 26th. Three markings could be seen, E5, F6 and G7, which I later named as the Sea of Columbus (E5), the Sea of Vespucci (F6) and the Sea of Galileo (G7). The longest of these, F6, occupied the middle part of the crescent disk of the lit hemisphere turned towards us. The two on either side, E5 and G7, were raised less above the terminator, so that only their northern tips could be seen and nothing at all of the intervening Straits of Cortez and Cassini, as they were completely within the dark hemisphere. In the Sea of Galileo, G7, the middle portion appeared darker than the remaining area of that marking and also of the others nearest to it. The tips therefore of those seas flanking F6 occupied almost the half-way positions on either side between the middle of the disk and the cusps, being actually slightly nearer to the centre of the disk than to the cusps.

"I fitted a micrometer to the telescope, the spaces between the threads of which, 63 in number, together are equal to two unciae of a Roman palm. Therefore a Roman palm contains 378 of these divisions. Now since the focal length of Campani's object-glass which was being used that day was 88 Roman palms, that whole of this length was equivalent to 33, 264 of the micrometer spaces. The diameter of Venus' globe XZ filled 6 of these micrometer spaces. Therefore in the same proportion of 33,264 to 100,000, 6 to $18^1/_2$ gives an arc of 0 degrees, 0 minutes, 45 seconds as the size of the angle subtended by Venus as seen from the Earth.

"On the diameter XZ from the centre of the marking F to the tip of the cusp X is three micrometer divisions, and the same number from that central point F to the other cusp Z. But the tips of the seas E and G are just under $1^1/_2$ micrometer spaces from the centre F, and a little more than $1^1/_2$ from the cusps X and Z. We saw this very clearly for almost a whole hour after sunset, that is from 5.25 pm. to 6.15 pm. From that site at the wooden bridge joining the Palazzo Barberi to the flat area of the gardens it was not possible to observe any longer, because the Earth's rotation carried the planet to a part of the sky obscured by the actual palace of Barberi.

"Three hours after the middle of the first observation, that is 8.40 pm., Venus could be observed conveniently from within the palace itself, in a very spacious hall looking out at the front, over 120 palms long and with a double row of windows facing west, where Venus was seen sinking towards the horizon. No longer from an open-air site but from an indoor location was the observation thus continued.

"So I set up the support of my object-glass near the large windows in that hall of the Palazzo Barberi remarkable for its great size and the adornment of the paintings by Sir Peter Beretti, on its arched ceiling. I was able to extend the thread to the correct distance of 88 palms with plenty of room left to position the eyepiece there and observe the planet conveniently for over half an hour before it set. For Venus was situated in Aries that day at 18 degrees 56 minutes about midday, according to the almanac, with a northerly latitude of 4 degrees 36 minutes. Its declination was 11 degrees 42 minutes after it crossed the meridian at 2.35 pm., and it was above the horizon at Rome right until 9.30 pm. That day the air was entirely free from haze, and in the complete absence of any disturbance due to wind the image of the planet was exceptionally clear. Also since we were within the walls of the hall and the observation was free from any but the tiniest vibration caused by the wind if there had been any, no part of the apparatus holding up the object-glass quivered appreciably, and we had a very clear view of these same markings observed three hours previously on the crescent disk of Venus, and we noted that they had not moved noticeably from their former situation, but that the larger one in the middle was still at the centre, as before. Also the two on either side whose tips projected over the terminator appeared just as at 5.30 pm. So when we compared carefully the disk drawing made at 5.45 pm. with that of 8.30 to 9 pm., we could find hardly any difference in the positions of the markings."

VI. Therefore since over one eighth of a revolution would have taken place within three hours if a whole rotation were completed within 23 hours, it followed that the centre of the larger marking F (Table II, February 26th), which was in the middle of the crescent disk of Venus at 5.45 pm., should have moved almost 50 degrees toward the cusp Z by 8.45 pm., and should appear to us to be beyond the position occupied by the marking at the side, E, at 5.45 pm., and the marking E should also have moved close to the cusp Z, to be almost vanishing from sight beyond the curve of the globe near Z. On the other hand, the marking G should have changed its position to the middle of the planet's crescent at F by almost 50 degrees in the three hours, and its forward movement through rotation should have been obvious if Venus completed a full revolution round its own axis inside 23 hours. Therefore the three markings which, at 5.45 pm., were spread over two quadrants of the globe XF and FZ equally, should have appeared compressed into the quadrant FZ with no part of them remaining in the quadrant FX, if the theory of a full revolution in 23 hours were correct.

But by a clear experiment we saw whenever we turned our eyes toward the planet in that aforementioned hall from 8.30 to 9 pm. with the same telescope of 88 palms, that the marking F6 was around the middle of the crescent, and it was almost the same distance measured by the micrometer from the tip E to the cusp X as it was from the tip G to the cusp Z, just as had been recorded at 5.30 pm.

It was therefore necessary to conclude that in the three-hour interval the Venusian markings had not advanced along their parallel more than two degrees of its circumference, and a daily progression of about 15 degrees in 24 hours could not give any appreciable change in 3 hours, which would be an eighth of this (under 2 degrees), whereas the theory of a full revolution within 23 hours would have shown an obvious movement of 47 degrees in three hours on the crescent of Venus, easily seen by observers with these big telescopes even without the help of a micrometer.

It is possible to calculate therefore, by comparison of that day's observations with the situation of the markings seen on preceding days, February 14th, 16th, 19th and 20th, at the same time of evening twilight, that the amount of daily progress is such that a quarter of the entire revolution is accomplished in about 6 days. Compare the diagrams of the observation of

February 14th, when the northern tip of the third sea C3 which we named after King Emmanuel, was about 30 degrees from the middle, with the diagrams of the observations made on February 16th and 18th. On February 16th the tip 3 was about the centre of Venus' disk. On February 18th that same tip was around the middle of the quadrant LZ. You will see that its daily progress is about 15 degrees, since in 4 days it had moved forward by about 60 degrees.

Finally, by comparing the observations of February 9th with others 24 days later, namely on March 5th, when the marking A or Royal Sea of King John V had almost come back again to the same part of the disk, and the nearby Sea of Infante Henry B2 was likewise returning almost to the original place occupied on February 9th, it should be obvious that in this period of 24 days a complete revolution takes place, or possibly in 24 days with the addition of a few hours.

VII. I have gradually worked out from further observations more removed in time from the first one how many hours should be added to the 24 complete days to fulfil a whole revolution.

Since proof that a revolution was complete had to be sought from the return of the markings to that part of the planet's disk where they had been seen first, and I understood fully that, through the parallelism of the axis maintained under the condition that we have explained, the curved paths of the markings would be inclined at an angle to the different sections of the orbit which varied day by day as regards the terminator, and I knew likewise that in the hemisphere being gradually turned towards the Earth the plane of illumination itself varied, encompassing first one region and then another, to avoid making erroneous guesses about the return of the markings, I thought it necessary to consult the diagram on Table IV and find out from it what the aspect would be at a given time of the hemisphere of the planet first seen on February 9th when Venus was at D.

I considered that one revolution would take from February 9th to March 5th (Table IV, Figure 1), as this amounted to 24 days and some hours not yet determined. However, since Venus travels from D to A in 56 days, and from A to B in the same number, completing half of its ellipse round the Sun in 112 days , eight days later five of the periods of 24 days would be completed. Therefore if it were at D on February 9th, it would be at B on June 1st, and eight days later, on June 9th, five periods of 24 days would be completed. Therefore if one revolution takes a few hours more than 24 days, the return of the markings to the situation noted previously on the disk should be expected one or two days after June 9th. The fourth return, however, would precede June 1st by about 15 days, that is to say May 17th or 18th, 24 days before the fifth rotation.

Now the situation of the markings carried on February 9th to the point L on the planet's globe and extending to LQ, that is the First Sea, by describing its parallel rL around the axis of revolution gZ, caused that same marking to emerge and rise, as it were, towards LQ through the hemisphere lit by the Sun. When Venus reached B on June 1st, this same marking was caused to depart from the lit hemisphere INÆML to the dark LQKI, that is on 17th or 18th May when Venus was on the curve ΘB of its orbit, and also on June 10th or 11th when Venus had progressed along the curve BI of its orbit towards V.

It was obvious that the marking of the First or Royal Sea, whose western limit the February 9th observation had shown at the centre of the hemisphere turned towards us, watching at Σ, would reach the same position on the globe at the end of the fourth revolution between Θ and B, or at the end of the fifth between B and V. The shape of that marking which we saw at LQ while Venus was at D, would also be visible between ÆNBL when Venus had reached B to observers looking at it from P.

The changed position should be assessed on the diagram in Table IV. We have said that the figure there should be regarded as a planisphere drawn for the plane of the Ecliptic with the spectator situated at the north pole of the Ecliptic, and therefore looking at it at right angles. On it is shown a general section of the plane of the Ecliptic with the globe of Venus travelling on its eight-month orbit ABCD round the Sun S. So the points A, B, C, D which in that scheme represent the centre of that planet, will also indicate, on the hemisphere of that globe projecting above the Ecliptic's plane, points on the globe's surface lying beneath the pole of the Ecliptic which acts as the zenith of their revolutions. If lines are drawn through those points and through Venus' axis of rotation gDZ, they will be at right angles to the equinoctial line NQ and serve the same purpose as circles of longitude and meridians on normal globes. Therefore the degrees on the Equator of Venus NQ equally extended to those meridian circles through the globe's spinning round on its axis will afford a measure and indication of a complete revolution. For on the Equator of Venus we have set the east and west boundaries of the markings or seas which clothe that planet. Therefore their approach to the line of that meridian through the North Pole of the Ecliptic, in which the zenith of the hemisphere of Venus turned to us, always observing from the plane of the Ecliptic, is necessarily found, will be a sure mark of the completion of a full revolution. What we have said about the zenith of the globe of Venus facing the north pole of the Ecliptic should also be applied to the nadir point of that globe diametrically opposite facing the Ecliptic's other pole, the South one. The start of a revolution can be fixed with reference to either.

It is possible not only to fix but also to observe when the individual degrees of the Equator reach the meridian line through the poles of the Ecliptic and Venus' axis of rotation, and when the eastern and western boundaries of the markings on that same equator already described by us reach this line one by one. For the centre of the disk of Venus turned towards us corresponds always to a point on that equator distant 90 degrees from the points on the meridians at the zenith and nadir of Venus. Thus when I saw, for example, on February 19th 1726 in the first of our observations the western edge of the First or Royal Sea at the centre of the disk of Venus turned towards us, the longitude of which, measured from its eastern edge, is 55 degrees, it necessarily follows that the upper line of meridian through the zenith of Venus towards the North Pole of the Ecliptic is 145 degrees, and the lower line of meridian through the nadir of Venus towards the South Pole of the Ecliptic is 325 degrees.

For the sake of brevity and clarity we will say in future that the first degree of Venus' Equator, for example, is at the nadir when the eastern boundary of the First Sea, from which we start counting the degrees of longitude, is on the meridian line near the South Pole of the Ecliptic. We will say that it is at the zenith when that same edge of the First or Royal Sea touches the top meridian line towards the North Pole of the Ecliptic.

VIII. After formulating these rules for establishing the beginnings and ends of revolutions, and their repetitions, it has not proved difficult for us to work out the number of hours to be added to the 24 full days found necessary in our previous observations to complete a revolution.

We compared firstly the beginning of a revolution counted from the first day of our observations, February 9th 1726, with the fifth revolution after that, about June 15th. The centre of Venus' disk facing us on February 9th was occupied by the western edge of the First or Royal Sea, whose longitude we said was 55 degrees. Therefore the line of the nadir was 325° longitude, 90° distant from this while the line of the zenith was 145°, the same distance in the other direction. The time for five revolutions (for which 24 complete days with the addition of some hours is needed), required an interval of 120 days when the sum of the complete days is made, and perhaps one day more for the total of extra hours not yet determined. Therefore I thought it necessary to turn my telescope on Venus at dawn on the days from June 7th to 11th when circumstances permitted. It was then at its morning elongation, and called Phosphorus. Although I was prevented from observing on June 9th and 10th, I succeeded on June 11th. On that day the western edge of the First or Royal Sea was not far from the centre of the Venusian disk turned towards us, but had moved a few degrees past it. There instead could be seen the eastern edge of the Second Sea of Infante Henry at about

70 degrees longitude. Therefore Venus had completed its fifth revolution round its axis since February 9th some time on June 10th. Thus we estimated that about five hours should be added to the 24 days for the period of one revolution.

Having reached this conclusion based on five revolutions within the same year, 1726, I thought it necessary to wait for a chance to compare another series of revolutions in 1727 to arrive at a more accurate estimation of the hours after a longer interval of time.

Since I had worked out that five revolutions are completed in 121 or $121^1/_2$ days it followed that in the 365 days of the civil year, about another fifteen would take place. Therefore we renewed our observations around July and August, when, as Hesperus, the planet was again approaching the Earth. And indeed the July observations were suitable to study the movements of the markings situated near Venus' equator on their curves to zenith and nadir, since in that position the planet turned its North Pole to us so that it appeared near the centre of the disk facing us, when we compared the longitude of the Northern Sea with the longitude of the others around Venus' equatorial zone. Towards the end of August and the beginning of September the equatorial zone and the markings situated on it that were lit by the Sun were presented to us in such a way that we were able to plot their positions and define their shapes at least for a quarter or half an hour in the evening twilight. A clear view was not permitted to us for longer than that, because the planet quickly approached the horizon as the Ecliptic did not ascend steeply. But our previous knowledge of the shapes of these same markings seen very clearly in February 1726, when Venus remained above the horizon for almost four hours after sunset, made it easy for us to recognize them again, so that in my latest observations I often made a diagram of what the markings would be like which we had to observe on that day, before we turned our telescopes towards the planet.

It was convenient, therefore, over a set interval of time from July 6th, when the far boundaries of the Sea of Marco Polo were situated on the line of the zenith and nadir, to the days in August and September when the markings in the equatorial zone were approaching these same positions, to define the longitudes of both the edges of the Northern Sea according to the times of their next revolutions, and to establish the farthest part from the north of that same sea, in our map of Celidography marked with the letter δ, which is almost the same circle of longitude as the centre of the First or Royal Sea, or a point just preceding this, about 20°. The other end of this Sea corresponds to the longitude of the west edge of the Fifth Sea, named after Columbus, at about 255°.

From both sets of markings, then, the northern ones seen in July and those of the equatorial zone observed in August and September, it was discovered that on about July 5th 1727, which is 511 days from February 9th 1726, Venus had completed 21 revolutions round its own axis. Therefore the amount for a single revolution should be fixed at 24 days and about a third of a day, or seven or eight hours. Rounding this off we assume that it is eight hours.

However, if one revolution of the globe takes 24 whole days and 8 hours, the daily movement will be 14 degrees, 47 minutes, 40 seconds, 16 thirds, worked out by dividing 360 degrees by 73 thirds of days, which is the total of 24 days 8 hours.

We knew that the 21st revolution was completed on July 4th 1727 from the observation we made three days later, in the evening twilight on July 7th 1727, when we saw the edge of the Northern Sea of Marco Polo, δ, which shares the same or slightly less longitude as the centre of the First or Royal Sea, that is, 20 degrees approximately, situated at the nadir, as can clearly be seen from the diagram of the observations. Therefore three days earlier, at evening on July 4th 1727, the longitude of that same nadir point would be 325° (since three days' rotation would move the Venusian markings through 45 degrees), that is to say the same degree of longitude occupied by that point in the evening observation of February 9th. So from February 9th 1726 to July 4th 1727 there were 21 complete revolutions of Venus round its own axis in 510 days, giving a value of 24 and a third days, or 8 hours, for one single revolution.

IX. But a more precise measure of one revolution could be defined more accurately if we were to wait for the end of the eight-year period, when Venus' movements and phases are repeated to observers on Earth on almost the same day of the civil year. Then from February 9th, 1734, as in 1726, an opportunity will be given to observers of the planet at its evening elongation, when it will be conspicuous above our horizon for almost four hours after sunset, to make a long inspection of the markings describing these same circles again in close proximity. So the east and west boundaries of the seven seas occupying the zone each side of the Equator can be compared daily, and the degrees of longitude at zenith and nadir worked out every day, thus establishing the end of a complete revolution. Comparison of these results with ours will enable a measurement of that period to be made accurate not only in days and hours, but also to the nearest minute.

X. Meanwhile, it is permissible to assume it is well established, in that it is very near to the truth and can be left to others to refine further, that 24 and a third days can be taken as the amount from our observations to date. That is to say, three revolutions take 73 days. This nicely fits our count of 21 revolutions from February 9th 1726 to July 5th 1727. Furthermore, it is sufficient to allow the construction of tables showing the cycles of revolutions. Since 21 revolutions amount to seven times three, and three revolutions give a time of 73 days, if the first and second revolutions are counted as 24 days and the third one as 25 (just as every fourth year an extra day is added to the Julian calendar to give 366 days, as opposed to the other three years which have only 365), a table can be made of these building blocks which will suffice to determine the positions of the markings and their aspect on the Venusian disk daily presented to us during the cycle of 21 revolutions. Then from this smaller cycle another larger one can be constructed applicable to a whole 8-year period, the start of the revolutions being two days earlier at the end of the eight Julian years, when 120 similar revolutions have been completed. For in a Julian 8-year period there are 2922 days. Five cycles of the smaller set of 21 revolutions give a total of 2555 days, which, when subtracted from 2922, leaves 367 days, which must be divided by 73, the time of 3 revolutions. The number 73 multiplied by 5 gives the total of days in the normal year 365, which subtracted from 367 gives 2 days remaining since the start of Venus' next revolution round its own axis after the 121 revolutions of the Julian 8-year period.

If this calculation is correct, Venus will show the same aspect to observers from the Earth on February 7th 1734 as it showed on February 9th 1726. But if the positions of the markings that I have described are restored a day earlier or a day later, that difference in time divided between the 120 revolutions will either increase or decrease the value of 8 hours allotted to the days of each one.

Sufficient has been said about the period of revolution. This spinning motion around the planet's own axis we have decided to call Perieilesis, using the Greek language (Περιειλησις). We will let the reader be the judge of the validity of the observations which determined the period as $24^{1}/_{3}$ days. The observations however which encouraged the belief in the very swift rotation period for the planet of 23 hours can be consulted by the reader when the advocates of that theory are mentioned in the scholarly letter from the Very Reverend P. Melchior de Briga, which we append at the conclusion of this small treatise.

Meanwhile we will proceed to tell the history of the third phenomenon of the planet Venus that we discovered: its tendency to maintain the parallelism of its axis in describing its eight-month orbit round the Sun.

CHAPTER VI

CONCERNING THE PARALLELISM OF THE AXIS OF ROTATION WHICH VENUS ALWAYS MAINTAINS IN ITS ORBIT ROUND THE SUN.

Summary of the Chapter.

I. It is deduced from observations that the alignment of the axis lies in the same direction whatever part of its eight-month orbit Venus may occupy, particularly by comparing different quadrants of the orbit in turn.

II. This is also shown from examining both poles visible alternately for half of the elliptical orbit.

I. The third discovery I made from my observations of the Morning and Evening Star is that the alignment of the axis of rotation lies in the same direction throughout every part of its eight-month orbit round the Sun. We have seen that the plane through the axis of rotation and the centre of the Sun cuts the Ecliptic at the 20th degree of Leo and Aquarius, in such a way that the North Pole of Venus, being raised about 15° above the plane of the Ecliptic, points straight at the constellation of Equuleus and the stars $\propto$ and β in the horse's head which are situated around that latitude and longitude, while the South Pole points to a place in the starry sphere a little below the heart of Hydra, at the 20th degree of Leo but with a latitude corresponding to the 15 or 20 degrees below. Therefore the axis of rotation of Venus always points in this direction, in whatever part of its orbit the planet may be.

All the observations show this parallelism, but it is especially obvious when we compare observations made on separate occasions when Venus was at parts of its orbit a whole quadrant removed from each other. As we have said, it takes Venus 56 days to traverse a quadrant. For those observations should be considered which we made towards the end of February, when the circles of rotation described by the markings were almost parallel to the terminator, and the plane at right angles to those circles produced through the axis passed through the Sun; and they should be compared with other observations about 56 days later, from April 20th onwards, when the planes of the parallels described by the rotations fell at right angles to the plane of the terminator.

A line from the centre of the Sun to the centre of Venus is the axis of illumination or of the circle of the terminator, and in separate quadrants of its eight-month orbit round the Sun two lines from the Sun to the beginning of each quadrant form a right angle; so it follows that these circles of illumination at the start of each quadrant form planes which if produced cut each other at right angles. Therefore more parallel lines also which meet one of the planes mentioned above at right angles at the start of a quadrant will if transferred to the other end of that same quadrant be parallel to the plane there which lies at right angles to the first one. And if planes are set up at right angles to these parallel lines, they will be parallel to one of these planes but will meet the other at right angles. In Figure I on Table IV, since the line SR from the centre of the Sun S to the centre of Venus R is the axis of the circle of illumination of the globe of Venus IRL on March 1st, the plane IRL is perpendicular to the plane of that circle. If it is supposed that the centre of Venus' globe on that same day, March 1st, is not only in that part of the orbit where the planet's axis of rotation lies on the plane MRS through the centre of the Sun, but also is at the point K on the plane of the Ecliptic, it follows that any one point on Venus' Equator separated from the pole at K by 90° of that globe's greatest circle, as for example the three points IRL, circle the actual Equator of Venus as they rotate, and circle the terminator at the same time. Other points on the hemisphere IKRL turning round the same axis MK will describe circles parallel both to the terminator and the Equator of Venus. But when after 56 days Venus has moved through a quarter of its eight-month orbit from that position to Δ, that is on April 25th, the line SΔ from the centre of the Sun to the centre of Venus will form a right angle with the line SKRM from the Sun to the end of that quadrant at R and the plane of the terminator IRL will meet the plane of the terminator IΔL, if prolonged, at right angles. Therefore if two lines are drawn at each end of the quadrant, R and Δ, parallel to each other, through the axis of rotation of Venus everywhere shown as KM, it is necessary that this axis, which is perpendicular to the plane of illumination in the first point of the quadrant R, will be parallel to the other plane IΔL at the last point of the quadrant which is at right angles to it, and whatever planes are parallel to the one axis will likewise be at right angles to the other. So the rotation of Venus round its axis KM and the circles described during that same rotational movement by individual points on the planet's globe and the markings on it around the axis KM will be at one extreme of the quadrant R parallel to the terminator IRL but at Δ will fall at right angles to the terminator IΔL. The planes of the circles of rotation on the days immediately before and after will show similar angles.

Now let us compare the diagrams of observations made from February 23rd to March 5th and on other days near them, with other observations made 56 days later. The daily progress of the markings will show that in the part of the quadrant R that lies on the line SKRM the circles of rotation are parallel to the terminator, while at the other extreme of the quadrant bounded by the line SΔ it will demonstrate that the circles of rotation are perpendicular to the plane of illumination. It does not much matter that we made no observation on April 25th, for soon after at the beginning of May the observations taken then show, when due allowance is made, that the circles of rotation described by the markings on April 25th must have been at right angles to the terminator, while the circles described by these same markings were parallel to the terminator on March 1st.

What we have concluded here about the parallel disposition of the planes and the correspondence of the axis of rotation KM with the axis of solar illumination SR on the supposition that the axis KM is on the plane of the Ecliptic MKSV, should be applied (with suitable adjustments made) to indicate the difference of the inclination of the terminator IRL at R to the Equator of Venus ITL and to the circles parallel to this described by the rotation of the markings round the axis Zg, when the plane KM is raised 15 or 20 degrees above the plane of the Ecliptic KM at the northern end K. For we said that this was about the amount of elevation from an examination of the observations explained above in Chapter III, Sections VI and VII.

II. This parallelism of the axis is also shown by detecting the rotation of the South Pole of Venus after its conjunction with the Sun on April 6th, when it had progressed to the quadrant of its orbit AΔB and we were observing from the arc EP. The theory of spherical perspectives or the analemma applied to the globe of Venus as seen from the Earth, while at its morning elongation as it rose before the Sun in the quadrant ΔBV and bore the name Phosphorus in May and June 1726, showed that the South Pole of the planet g always appeared clearly to us, and the marking surrounding it which we have called the Sea of Magellan was constantly turned towards us while the centre of Venus was at the point Δ of its orbit and cut in two places by the terminator IΔL. But when the planet had moved to B it appeared clear of the terminator IBL along the arc Lg, while the other markings of the southern hemisphere from the South Pole g to the Equator NBQ described wider circles of rotation in proportion to their distance from the Pole, and gradually receded from the crescent of Venus visible to us and were hidden beyond the terminator, disappearing towards the north cusp of the planet and emerging from the south one.

This demonstration of the parallelism of Venus' axis of rotation was completed by an observation of the planet's North Pole and the turning of the Northern Circumpolar Sea (which we called after Marco Polo) which we made in 1727 from July 7th to July 18th. At that time the North Pole, marked Z in Table IV, and in the diagram of those observations on Table III marked with the letter S, was almost at the centre of the hemisphere turned towards us at the evening elongation from the Sun when it is called Hesperus. The position of the Pole and axis of rotation was made obvious from a comparison of the progress of that marking, as the accurately drawn diagram shows. For on July 7th that semicircular marking n o z p r stretched beyond the diameter KSM through both cusps and the centre of the hemisphere turned towards us by an equal amount at both of its ends n o and p r, and curved on its semicircular course towards the terminator KM, keeping the shape of the Latin letter ⊃ inverted in the drawing made at the telescope with a single lens eyepiece, as is customary[1]. But on July 10th the end of the marking p r had progressed beyond the plane KSM towards X, and the other end n o had receded from the plane through the cusps and the centre, KSM, towards V, covering the distance appropriate to three days' revolution (which we have established is $1/8$ of a rotation approximately). Finally on July 18th, 11 days after the first observation of July 7th, half a full revolution was almost complete; the circumpolar marking n o z p r was at the opposite position on Venus' disk and resembled the Latin letter C, thus showing that the Arctic Circle, so to speak, round the planet's pole was almost filled with the letter C⊃ in both of its positions. Thus the site of the North Pole was revealed to us, and a description of the northern region could be given. The end of the polar marking of Marco Polo designated n o is a little nearer to the Pole S than the other end marked p r, as we said earlier in Chapter V, Section VII.

The heliocentric motion of Venus had returned the planet on June 29th 1727 almost to the same part of its orbit that it had occupied on April 6th and November 17th 1726, namely to the letter A on Table IV. So when we were looking from a little way beyond F towards P at the Sun S on July 7th in the 15th degree of Cancer, and Venus at A in the direction of the line FA, making an angle of 39 degrees with SF, and referring to the 20th degree of Leo in the circle of the Zodiac, the part of the sky towards which the planet's axis of revolution constantly points, the North Pole also should have been conspicuous and indeed at the centre of the hemisphere of the planet turned towards us. Therefore the circle of revolution described by the polar sea of Marco Polo was exposed to us, and proved that the parallel disposition of the axis was maintained by the planet while pursuing its eight-month orbit round the Sun.

To ensure a satisfactory view of that circumpolar marking at a time when the Earth was further from Venus than the Sun, a very clear day had to be chosen. Such was July 7th, when we observed from the Alban Hill. After morning rain, the clarity was increased opportunely by a slight freshening

---

[1] *He means 'not having a terrestrial eyepiece' (which would have given an upright image); the two lenses in a Huygenian eyepiece act as a single lens and the image is inverted.* **P.F.**

of the Mistral in the afternoon, which completely cleared the evening sky of haze, particularly from our observing site high on the hill, with a clear horizon towards the Western Sea giving an extended view of Hesperus at the evening elongation. Similar too was our other observation of the planet from the Palatine Hill in Rome on July 18th, a day of equally good visibility. The length of the telescope made by Campani was 94 palms. The aperture of the object-glass was 3 unciae of a Roman palm[2]. The observation time was half an hour after sunset. I have given the circumstances of this observation on purpose, in case anyone wishes to repeat the experiment in the future under the same conditions at a similar elongation where the planet's position in the Zodiac ensures that it will be above the horizon for over two hours after sunset.

To find out when an opportunity will occur of expecting the same conditions in the next few years, the Ephemerides compiled by the renowned Eustace Manfred and others up to the year 1750 will give abundant reliable information. However to spare the reader trouble, we include the information in the last chapter of this short work. Now we will proceed to the observations of parallax.

---

[2] *2.2 inches Imperial measure, which would give a maximum possible resolution of 2 seconds, or about 1/10 th of the diameter of Venus (probably less due to chromatic aberration).* **P.F.**

CHAPTER VII

CONCERNING OUR VERY CAREFUL OBSERVATIONS OF THE PARALLAX OF VENUS AND THE CONCLUSIONS DRAWN FROM THEM.

Summary of the Chapter.

I. *The parallax of Venus against the background of the fixed stars was the fourth of our recent observations of the planet.*

II. *Cassini's method.*

III. *Comparing the difference in Right Ascension between Venus and a fixed star visible with it firstly on the Meridian and then beyond the Meridian.*

IV. *Observations on these lines when Venus was close to Cor Leonis or Regulus from July 1st 1716 to July 4th.*

V. *The conclusion from these that the horizontal parallax of Venus on July 3rd was 0 degrees, 0 minutes 24 seconds, 20 thirds.*

VI. *From this determination of the parallax of Venus its distance from the Earth is worked out.*

VII. *This also enables the Sun's distance from the Earth to be established.*

VIII. *Furthermore the size and dimensions of the whole Solar System can be determined.*

IX. *It was not possible to repeat the same experiment in 1724, which would have greatly supported the confirmation of the first experiment.*

X. *We have substituted for that particular repetition other observations even more thorough, but not quite as suitable for giving trustworthy results.*

XI. *It is important to repeat our former efforts using Sirius* and Spica in Virgo when Venus passes close to these fixed stars.*

---

*As Sirius is far south of the Ecliptic, it would seem that Bianchini means Regulus. **S.B.**

I. The last of the four recent discoveries revealed in this short work is the careful investigation of Venus' parallax, which we were able to undertake ten years ago. As we have said, this also enables the whole Solar System from the Sun to Saturn to be measured. Astronomers will pass their judgement on the method of the experiment and the results of our investigations after reviewing the following observations. Before describing them, I will bring up a few points for consideration.

The minute angle subtended by the Earth's diameter at the vast distance of this heavenly body from us is shown to be equal to that formed by two straight lines meeting at that body from two points on the Earth's globe separated by a quarter of its greatest circle drawn through the globe's centre, and is called by us horizontal parallax. As I say, that angle is so minute at such a great distance of these heavenly bodies that it evaded measurement by the old astronomers, who before the invention of the telescope were deprived of instruments adequate to detect quantities smaller than the naked eye could distinguish. So with the exception of the Moon, whose nearness ensures that the angle is almost a whole degree even at its average distance and thus is easily detectable, they met with complete failure regarding all the other planets. The difficulty of measuring so minute an angle clearly and accurately was increased by the equally difficult conditions necessary for the observations. For it seemed impossible to measure such an angle unless two observers separated on the Earth's surface by a quarter of the perimeter of the whole globe turned their attention at the same time to the planet, comet or other body beyond the Moon's orbit when it was visible to both of them.

II. Both of these difficulties were removed, however, when J.D. Cassini, undoubtedly the prince of modern astronomers, used the presence of the great comet which appeared in 1680 and 1681 to publish in his excellent treatise on that subject a method of determining that angle with great accuracy by one observer alone. It was completely reliable and convenient, and involved measuring the difference in R.A. between one of the fixed stars and the heavenly body whose parallax was being investigated, first as it crossed the meridian and then at other hourly circles, with a telescope fitted with a micrometer, and using a reliable clock for measuring the time with a swinging pendulum adequate to divide half-seconds at least, thus showing the difference in R.A. to an accuracy of at least an eighth of a minute.

This is a very reliable and accurate method, and easier than others to carry out in practice, but it can be applied more readily when the heavenly body whose parallax is being sought can be observed on the meridian when a fixed star is very close to it and compared six hours before or after they both cross the meridian. This often happens in the case of the Moon, Mars,

Jupiter and Saturn. But it did not seem an easy proposition in the case of Venus. For although Venus can be visible by day even with the naked eye, that is to say with no telescopic aid, not only at the meridian but beyond it, when the Sun is present, nevertheless the fixed stars that are visible moving in a straight line close to Venus are lost to our sight even with the use of a telescope, as long as the Sun is above the horizon.

III. Yet because I had noticed that fixed stars of the first magnitude could be excluded from this number, in that I had observed that Sirius could be seen clearly during daytime with a telescope of only three palms on June 29th and 30th as it reached the meridian together with the Sun, at about the 8th or 9th degree of Cancer, I was confident that this method could be adapted to Venus too, when it was near Regulus or Cor Leonis, so that they were both visible in the field of view together (even when the tube was extended to 23 palms). From astronomical tables and the calculations of ephemerides I knew that this conjunction of the star and the planet could be expected on July 3rd, 1716. Therefore I made my preparations for that observation two days before this.

IV. Shortly before sunset I turned a shorter telescope of 7 palms towards the planet Venus, and on the threads of the micrometer fitted to the telescope, near the parallel of Venus which preceded, I first awaited its approach, then that of Regulus following, to the hourly circle to find out the difference in right ascension and declination between Venus and Regulus. This was easy to record, since shortly before sunset Regulus too was clearly visible in the tube of that length. Then I turned Campani's longer telescope of 23 palms[1] on Venus, and shortly afterwards adjusted it to take in the transit of Regulus in its field of view, and tried to get a clear enough image of Regulus' light to define the star's position accurately even with the Sun not yet setting. Thus I hoped that even when Venus and Regulus crossed the meridian, the light of both would have sufficient strength in that same telescope of 23 palms to present a clear image to our eyes. To improve the sensitivity of the organ of sight I darkened the room where the meridian line was marked, and opened a small window there just wide enough to line up Regulus and Venus in the aforementioned telescope of 23 palms on the line of the meridian at the correct elevation, and I also attached a sextant to serve as a setting circle for the telescope, so that I could locate Venus, which was shortly due to approach the meridian line, in good time. I had made all these preparations for July 2nd and 3rd, when the sky was very clear. This is the series of observations, which was continuous from July 1st to 4th.

---

[1] *17 feet Imperial measure, set up in a darkened room and aligned with the meridian.* **P.F.**

At Rome it was the fourth day of the holidays on July 1st, 1716.
The sky was very clear.

Hours	Minutes	Seconds	After Noon
			The micrometer threads on the telescope were placed near the plane of the parallel of Venus, and others at right angles to these, showing as is customary the hourly circles in the same micrometer.
A 8	16	0	The limb of Venus preceding in the diurnal revolution and lit by the Sun reaches the hourly circle.
8	20	4	Regulus reaches the same hourly circle. The difference of declination between the centre of Venus' disk, which was north of Regulus, and Regulus which was further south with respect to Venus was according to the micrometer threads, 0 degrees 40 minutes 4 seconds. The difference of right ascension was 4 minutes 4 seconds, as shown on a portable clock as 582 oscillations.

The fifth Holiday, July 2nd. The sky was very clear.

B 3	5	6	Venus was seen near the meridian clearly with the telescope, then with the naked eye. But Regulus, which crossed the meridian after Venus, could not be seen at all with the telescope.
C 7	19	20	The preceding limb of Venus lit by the Sun reaches the micrometer thread of the hourly circle.
7	20	19	Regulus reaches the thread of the hourly circle.
D 7	23	20	The illuminated limb of Venus reaches the thread.
7	24	20	Regulus reaches the same thread.
E 7	33	20	The investigation in the difference of R.A. is repeated at the same time as an observation of the difference of declination between the centre of Venus and Regulus, which was found by micrometer to be half the Sun's diameter, or 0 degrees 15 minutes 50 seconds.

continued overleaf

Hours	Minutes	Seconds	After Noon
F 8	29	0	Venus was north of Regulus. In order to measure more accurately the difference in R.A, between the preceding limb of Venus lit by the Sun and Regulus, I placed my ear near a clock in a portable case made by Quare[2], whose 143 vibrations or oscillations equal 1 minute or 60 seconds, and can be heard and counted by an observer who concentrates his eyes on the micrometer and his ears on the clock. This method had become easy to me with experience and practice. So now from the approach of Venus' limb to the approach of Sirius[3] to the same thread of R.A, or hours I counted 137 oscillations of that clock, equalling 55 seconds. The difference in declination was observed to be 15 minutes 7 seconds.
G 9	0	0	The difference of R.A. between the preceding limb of Venus and Regulus is 121 clock oscillations, equalling 52 seconds in time. The difference in declination was observed to be 14 minutes 24 seconds.
H 9	20	0	The difference of R.A. was measured in seconds with another oscillatory clock made by Thuret[4], with a longer pendulum, by the same method and found to be 52 seconds, equalling 115 oscillations on the portable clock. The difference in declination was 14 minutes 0 seconds.

---

[2] *Daniel Quare 1647-1724, worked in London, made portable clocks and barometers, also repeater watches, famous for his skill.* **P.F.**

[3] *Regulus rather than Sirius is obviously meant.* **S.B.**

[4] *There were two Thurets, Isaac and Jacques, working in France.* **P.F.**

The Sixth Holiday, July 3rd, a very clear sky.

Hours	Minutes	Seconds	After Noon
K			V. At three o'clock in the afternoon as Venus was approaching the meridian, I pointed at it the optic tube of 23 palms made by Campani, and since Regulus preceded it on almost the same parallel by a minute and a half, the atmosphere was very steady, I succeeded in seeing clearly in that telescope Regulus preceding, and managed to make an accurate note of the difference in right ascension and declination between it and the illuminated limb of Venus following. So from 3 pm. to 3.10 I repeated the experiment, until Regulus just touched as it departed the daily parallel marked by the micrometer threads,
K 3	10	0	and I counted 218 oscillations of the portable clock from the contact of Regulus which preceded to the contact of the illuminated limb of Venus at the same thread of the hourly circle. 143 of these oscillations are equal to 60 seconds or a minute. So the difference in right ascension at this moment in time when Venus had passed the meridian about 6 minutes earlier was 218 oscillations of the portable clock, which corresponds to 1 minute, 31 seconds and 30 thirds approximately. There was a difference in declination of one of the micrometer divisions, 67 of which are subtended by the Sun's diameter; or expressed in degrees of the great circle, 0 degrees, 0 minutes, 28 seconds. So Regulus was 28 seconds north of the centre of Venus.
L 6	56	0	The difference in right ascension between Regulus preceding and the illuminated limb of Venus following was 275 oscillations of *continued overleaf*

Hours	Minutes	Seconds	After Noon
M 7	14	0	the portable clock, which is equivalent to 1 minute, 55 seconds, 30 thirds. The difference in declination was $9^1/_2$ micrometer divisions, of which the Sun's diameter subtends 67. So of the great circle this is 0 degrees, 4 minutes, 31 seconds, by which amount Regulus was north of the centre of Venus. The difference in right ascension is 283 clock oscillations, and the difference in declination 0 degrees, 4 minutes, 50 seconds.
N 8	0	0	The difference in right ascension is 304 clock oscillations. The difference in declination is 5 minutes, 40 seconds.
O 8	29	0	The difference in R.A. is 313 oscillations, equivalent to 2 minutes, 11 seconds.
P 9	0	0	The difference in right ascension between Regulus and the limb of Venus, as mentioned above, was 322 clock oscillations, which is 2 minutes, 15 seconds.
R 9	10	0	The difference in right ascension is now 324 clock oscillations, which is 2 minutes, 16 seconds. The difference in declination is $14^1/_2$ micrometer divisions, or 0 degrees, 7 minutes, 6 seconds of the great circle.

The Sabbath Day, July 4th. The sky was very clear.

Hours	Minutes	Seconds	After Noon
T 8	25	0	The difference in right ascension between Regulus and the illuminated limb of Venus, as mentioned above, equals 5 minutes, 10 seconds, as it was 739 oscillations of the usual portable clock.
V 8	33	0	The difference of right ascension is found to be almost the same.
			From these observations, therefore, it is deduced that the horizontal parallax of Venus on July 3rd was 0 minutes, 24 seconds, and 20 thirds of the great circle, through the following calculations and comparisons.

From July 1st to July 2nd.

From the 1st to the 2nd of July, according to the observations A and F separated by 24 hours and 13 minutes, the right ascension of Venus altered by 3 minutes 9 seconds, corresponding to 449 oscillations of the portable clock (see letter F above). Therefore within 24 hours, 0 minutes, the right ascension changed by 3 minutes, 5 seconds, which is 441 oscillations of the portable clock[5].

From July 2nd to July 3rd.

According to the observations G and P separated by 24 hours 0 minutes, Venus was found to have changed its right ascension with respect to Regulus by 443 oscillations of the portable clock.

From July 2nd to July 4th.

According to the observations F and T from 8.29 pm. on July 2nd to 8.25 pm. on July 4th, the right ascension of Venus altered by 874[6] oscillations of the portable clock. Therefore within 24 hours it changed by 437 oscillations approximately.

---

[5] *189 sec in 1453 min ought to give 187.3 sec in 1440 min i.e. 3 min 7 sec if this is simple proportion (0.9% reduction)?* **P.F.**

[6] *876?* **P.F.**

From July 3rd to July 4th.

According to the observations D and T V from 8.29 pm. on July 3rd to 8.29 pm. on July 4th, the change in right ascension was 440 oscillations of the portable clock, and this had been found to be 441 from July 1st to 2nd. and on the other days 437 and 443, the average of which gives 440 oscillations within 24 hours for July 3rd.

If the change in right ascension within 24 hours is 440, a quarter of that time, namely 6 hours, will give a change of 110, if Venus exhibits no parallax.

The observation K on July 3rd at 3.10 pm., when Venus and Regulus six minutes earlier were crossing the meridian was six hours removed from the observation R at 9.10 pm. For the difference in the observations was as follows:

Observation K	218
Observation R	324
	106
Without parallax it should have been	110
Therefore the parallax was	4 vibrations.

Now 4 vibrations of the portable clock converted to minutes of time gives 0 minutes, 1 second and 40 thirds, since 143 oscillations are equal to 60 seconds. On the daily parallel of Regulus and Venus, 0 minutes, 1 second and 40 thirds subtends an angle of 0 degrees, 0 minutes, 25 seconds and 0 thirds.

In order to relate that angle on the parallel of Regulus and Venus, which on July 3rd subtended an angle of 25 seconds, to the arc of the great circle, the declination of Regulus and Venus must be taken into consideration. On that day it was 13 degrees, 21 minutes. Therefore the amount to complete a quadrant of the circle is 76 degrees, 39 minutes.

This amount of 76 degrees 39 minutes, if a complete sine is regarded as 100,000, will on that scale be 97,297. Therefore an arc of 25 seconds 0 thirds on the parallel of Venus will correspond to an arc of 24 seconds 20 thirds on the Equinoctial or any other great circle of the sphere. That will be the amount of the horizontal parallax of Venus, or the angle beneath which half the Earth's diameter, subtending an angle of 90 degrees from the meridian to the circle of the sixth hour, was seen from the observation K at 3.10 pm. to the observation R at 9.10 pm., and that was the object of our enquiry.

So the conclusion drawn from these observations is that the horizontal parallax of Venus is 0 degrees, 0 minutes, 24 seconds and 20 thirds.

VI. From the parallax of Venus, or the angle which half the Earth's diameter would subtend if seen from Venus, it is established that on July 3rd 1716 the distance between that planet and ourselves was equal to 8,000 semi-diameters of the Earth. For as the tangent $12^{1}/_{2}$ of the aforementioned angle of 0 degrees, 0 minutes, 24 seconds and 20 thirds is to the complete sine of 100,000, so also is one single semi-diameter of the Earth to the 8,000 semi-diameters of the Earth, the distance of Venus on that day from the terrestrial globe. Finally, once the parallax of Venus and its distance is known, we can proceed to calculate the Sun's distance and parallax.

VII. On August 28th of that year 1716 it happened that Venus and the Sun, as seen from the Earth, were in the 6th degree of Virgo, as is established from the Tables of movements of celestial bodies and the calculations of Ephemerides. Therefore Venus as seen from the Sun was in the 6th degree of Pisces. The observation of parallax undertaken on July 3rd was 56 days before this conjunction, during which time Venus completes a quarter of its eight-month orbit round the Sun. So Venus as seen from the Sun was in the 6th degree of Sagittarius on July 3rd. We saw the Sun in the 12th degree of Cancer on that day, and the Earth as seen from the Sun was in the 12th degree of Capricorn, and Venus along with Regulus was in the 26th degree of Leo as seen from the Earth.

So on July 3rd, in the triangle STV formed by the lines TS, the distance from the Earth T (Terra) to the Sun (S); TV, the distance from the Earth to Venus; and VS the distance of Venus from the Sun, the angle S was 36°, the angle T was 44° and the angle V was 100°.

Therefore, as the sine of the angle S, 36°, which is 5878, is to the sine of the angle V, 100°, which is 9848, so is the distance TV of Venus, 8000 semi-diameters of the Earth to the distance of the Sun from the Earth TS, that is

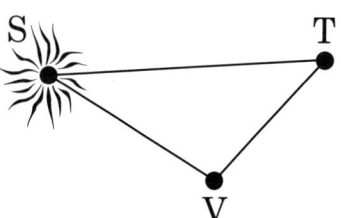

13,403 of these half-diameters, or in round numbers, 13,400. The Sun's parallax will also be found from this same distance. For as 13,400 is to 8,000, so the horizontal parallax of Venus on July 2nd, found above to be 0 degrees, 0 minutes, 24 seconds and 20 thirds, or 1440 thirds, is to 858 thirds, which gives 14 seconds, 18 thirds as the Sun's horizontal parallax[7].

VIII. Finally from the knowledge of the Sun's distance from the Earth gained from the observations of Venus, we proceed easily to the distance of Jupiter and Saturn from the Sun, and indeed to the dimensions of the whole

---

[7] *Accepted value today = 8.8 seconds. Bianchini was only out by a factor of 2 in his determination of such a small angle - a remarkable achievement, considering the instruments he had to use! (His value would put the Sun's distance at about 50 million miles as compared with the accepted 93 million, or the value obtained by Cassini using Mars in 1672, 84 million.)* **P.F.**

Solar System. For Jupiter's distance from the Sun is shown to be 5 times the distance of the Sun from the Earth, while Saturn is 10 times further from the Sun than the terrestrial globe is, as is proved from the eclipses of the satellites of these planets.

This will solve therefore a most pressing problem in Cosmography, Astronomy and Physics, namely the size of the whole Solar System, which follows as a direct corollary from the observation of Venus' parallax, and is fixed so finely and accurately by this method that we can hardly expect equal certainty, it seems, from any other observation undertaken hitherto.

IX. I was very keen to repeat this experiment 8 years later, when Venus was in fact close to Regulus again on July 3rd 1724 indeed in such proximity that the star and planet would almost seem to be touching. But I could not obtain permission in 1724 to place my telescope of 23 palms in that darkened room in the Palatine building which I had used during the reign of Pope Clement XI in 1716, when I opened a small window in it to conduct my observations. For Pope Clement, who encouraged my research like a true Maecenas, had now died, and no other suitable place for a new observation was available in any premises I could hire. Thus I had to abandon my attempt to repeat such an important experiment.

X. I tried to make up for the fact that I could not confirm in 1724 my observation of the difference in right ascension of Venus when seen in close proximity to Regulus, by a measurement of a similar difference but one which would require a much more laborious procedure in 1727 on September 19th when Venus and Saturn, although separated by several Zodiac signs, were however in the same daily parallel, at an equal declination of 19° south of the Equator.

This situation, whereby both planets were at the same declination but separated by several signs, meant that the differences in right ascension of Venus and Saturn could be recorded several hours before they crossed the meridian. So by comparing both angles, those taken before and after the meridian crossing, I mean, the measurement would have a much greater degree of accuracy than if the difference were only recorded over a period of 6 hours after the meridian crossing (as happened with Regulus in 1716).

The difficulty remained, however, of measuring to the desired degree of accuracy by means of the clock's pendulum, which would be required to maintain a very steady rate throughout the intervening time from the departure of Venus to the contact of Saturn at the micrometer threads indicating the same hourly circle.

Nevertheless, this steadiness is confirmed first by the theory of pendulum oscillations, propounded by the great Galileo and developed by succeeding professors of mathematics, and ingeniously adapted to clocks by skilled craftsmen, and secondly by a series of experiments which I performed almost daily as Sirius crossed the meridian, assuring me of the absolute regularity of clocks of this type fitted with the longer pendulum. After placing on a roof an iron sheet at right angles to the horizon, I go onto a balcony more than 60 feet away from the sheet, oblique to the plane of the meridian and turned towards the north relative to the sheet. Here I observe Sirius through this stable "cross-staff" fixed 60 feet away with a 3-palm telescope on an iron stand firmly fixed to the wall, and I wait until it touches the sheet, likewise fixed to a wall. This I call the contact of Sirius with the sheet, and after this contact I count 8 minutes and 24 seconds before Sirius reaches the circle of the meridian.

Since it is possible on every clear day to see Sirius exactly at the moment of this contact, I can, whenever I wish, test the regularity of the clock by observing Sirius passing behind the sheet for several consecutive days and noting the hour, minutes and seconds as recorded by the clock day by day, to see if the number of pendulum oscillations stays the same, or gradually increases or diminishes. Thus I was finally reassured about the reliability of the clock and its pendulum and the regularity of its timing.

After establishing this reliability over a period of 24 whole hours, I feel I can trust the clock and its pendulum to indicate the individual parts of that period accurately, which I have applied to the following observations of the contact of Venus and Saturn and also some of the fixed stars near Saturn, in order to confirm my previous measurement of the parallax of Venus.

I placed several telescopes of 3, 5, 7 and 11 palms, each fitted with micrometers, in various parts of that room from which the sky was visible through the windows, so that I could conveniently observe the approach of Venus in the daytime and Saturn at night to the various hourly circles before and after the meridian. The telescopes were not moved from the time when Venus, which preceded, touched the micrometer threads set at the daily parallel at the hourly circle indicated in each telescope by its own micrometer, and at which Saturn would arrive $6^{2}/_{3}$ hours later, and I recorded the hours, minutes, and seconds by no other means than placing the clock fitted with the longer pendulum close to the telescopes, and using my eyes and ears to count the individual seconds of time.

I have described the arrangement of this equipment to demonstrate the reliability of our accurate measurements of the small intervals of time over the long-extended period of 6 hours by which the contact of Venus preceded the entry of Saturn and the fixed stars near it to the area of that same hourly circle.

After this exposition of methods follows the series of observations from which I will select the most successful.

Rome, September 18th, Fifth day of the Holiday.

For the occasion of a lunar occultation of Venus, which we would have observed today if clouds had not prevented it, several telescopes, each fitted with a micrometer, were set up, fixed on various hourly circles and with the micrometer threads suitably positioned near the daily parallel of Venus and Saturn. I will list six in particular.

Telescope 1. On the plane of the circle of 10 hours 14 minutes past midnight, or 1 hour 46 minutes before noon, forming an angle of 26 degrees 30 minutes with the meridian.

Telescope 2. At 10 hours 10 minutes past midnight, turned away from the meridian towards the east through an arc of 25 degrees 15 minutes.

Telescope 3. On the meridian.

Telescope 4. On the plane of the circle 2 hours 42 minutes past the meridian.

Telescope 5. On the plane 3 hours 6 minutes past the meridian.

Telescope 6. On the plane 3 hours 25 minutes past the meridian.

The clouds, which in the morning and at noon had prevented a view of the sky, had luckily receded by a few minutes past 2 pm. Since Venus had emerged shortly before from behind the Moon's disk, I was able to measure the difference in right ascension at 2.10 pm. between the limb of Venus lit by the Sun and preceding in the daily rotation, and the limb of the Moon similarly lit and preceding in the daily rotation but following the approach of Venus to the hourly circle. This was the result:

2h 10m 16s	The preceding limb of Venus touches the hourly circle on the micrometer of the 10-palm telescope.
2h 10m 28s	The preceding and illuminated limb of the Moon reaches the same hourly circle on the same micrometer.
2h 14m 59s	The same telescope and micrometer is readjusted to record the approach of Venus to the thread of the hourly circle, and the preceding limb reaches it.
2h 15m 19s	The preceding limb of the Moon, lit by the Sun, reaches the same micrometer thread.

The difference in declination of the northern limb of the Moon, from the nearest limb of Venus, which was further north than the Moon, was 12 micrometer divisions, of which the Sun's diameter today subtends 34.

I leave out the rest of this day's observations relating to a comparison of the Moon with Venus, since they are not relevant to the present inquiry examining the parallax of Venus. I will recount those which we undertook in connection with this inquiry concerning the difference in right ascension with Saturn and various fixed stars which appeared close to it both on and beyond the meridian, according to the method of Cassini. It will suffice to publish the more successful ones, and in the margin I will prefix each one with the letters A, B, C etc., so that they can all be readily referred to and compared with each other. It is appropriate to begin with those relating to the investigation of the clock's regular motion by a comparison of the numbers of hours and minutes with the daily revolution of the celestial sphere.

The motion of the clock fitted with the longer pendulum was never interrupted from July 15th to September 25th. On July 15th, when the penumbra[8] of the preceding limb of the Sun touched the meridian line, the pointers of our clock showed 0 hours, 0 minutes 54 seconds, and at the departure of the penumbra of the following solar limb they showed 0 hours 4 minutes 25 seconds. Therefore the centre of the Sun was on the meridian on July 15th when the clock registered 0 hours 2 minutes $39^1/_2$ seconds.

The clock measured a full revolution of the sphere in 23 hours 56 minutes 56 seconds. For the return of Sirius to the meridian was earlier every day by 3 minutes 4 seconds than the full 24 hours, as recorded by our clock in the following observations.

On August 13th Sirius touched the perpendicular sheet that we described previously, fixed on the roof at an oblique angle towards the east, 8 minutes 24 seconds before the meridian plane on the same parallel as Sirius, by which I mean it reached the sheet and hourly circle when the clock said
A 9 hours, 21 minutes, 0 seconds, whereas it reached it the next day, August 14th, when the clock said 9 hours, 17 minutes, 56 seconds.

The difference is 0 hours, 3 minutes, 4 seconds. Again on September 15th Sirius touched the sheet when the clock said 7 hours, 40 minutes, 19 seconds. Seven days later, on September 22nd, it touched the same sheet at 7 hours, 18 minutes, 40 seconds. The difference is 0 hours, 21 minutes, 30 seconds. Divided by 7 days, this gives a daily precession of 3 minutes, $4^2/_7$ seconds.

---

8 *The Sun's "shadow and half-shadow" mean the shadow cast by the gnomon of a sundial, or of an upright rod "on a line marked on the pavement". P.F*

Therefore on September 19th, the day we selected to investigate the parallax of Venus (since on that day the daily parallel of Venus and Saturn could be considered the same), the clock measured a complete revolution of the heavens, or the full 360 degrees of the Equator, in 23 hours, 56 minutes, 56 seconds. An hour on the Equator, that is to say 15 degrees, it registered as 0 hours, 59 minutes, 52 seconds and 20 thirds.

On the same principle by which the number of hours and minutes was investigated which the clock took to measure a full revolution with Sirius or another fixed star returning to the meridian, the time taken by the Sun, Venus and Saturn for a full revolution, or the return of each of these bodies to the meridian or any other hourly circle, had to be measured in hours and minutes by the clock on September 19th, in order to make an estimate of the difference in such a time and work out the parallax of Venus.

As regards the Sun's daily return to the meridian, this is recorded nearly every day when time allows. It was found on September 9th to consist of 24 hours and 34 seconds on our clock.

B For from the crossing of each limb of the Sun along with its penumbra through the meridian line marked on the pavement, the centre of the Sun was on the meridian as follows:

On. Sept. 19th when our clock registered	0h	47m	15s
On the following day, Sept. 20th,	0	47	49
The difference is	0	0	34
Added to this the difference in the precession of Sirius	0	3	4
The sum is	0	3	38

I wish to indicate the difference in right ascension of 0 degrees 54 minutes 30 seconds which should be added to the arc of 0 degrees, 58 minutes, 47 seconds of the daily progress of the Sun in the Ecliptic from September 19th, when its position at midday was in Virgo at 25° 53' 47" to midday on September 20th at 26° 52' 34" in the same sign.

A revolution of Venus to the same hourly circle on September 19th was completed in 23 hours, 59 minutes, 53 seconds on the clock, for it preceded the 24 whole hours of that clock by 7 seconds.

On September 10th, Venus was on the same daily parallel at almost the same declination as Sirius.

Sirius touched the plate which we fixed on the roof on the heavenly plane 8 minutes, 24 seconds before the meridian when the clock showed the hour as 7 am. 55 minutes, 48 seconds.

The centre of the Sun reached the meridian at 0 hours, 42 minutes, 50 seconds on the clock. The preceding limb of Venus reached the plate set up on the roof, which Sirius had reached in the morning, at 3h 19m 2s pm. on our clock.

On September 11th the centre of the Sun returned to the meridian when the clock registered 0 hours, 43 minutes, 3 seconds. The preceding limb of Venus reached the plate on the roof, as described above, at 3 hours, 19 minutes, 15 seconds on the clock.

On September 15th we observed a solar eclipse. On that same day, the centre of the Sun touched the meridian as the clock showed 0 hours, 45 minutes, 19 seconds.

On September 17th Sirius reached the plate on the roof at 7 hours, 34 minutes, 18 seconds by the clock.

The centre of the Sun touched the meridian at 0 hours, 46 minutes, 16 seconds.

On September 19th the centre of the Sun was on the meridian at 0 hours, 47 minutes, 15 seconds.

C On the same day the preceding limb of Venus appeared in the sixth telescope set up at the hourly circle 6 hours 25 minutes removed from the meridian towards the west when the clock showed 6 hours, 59 minutes, 47 seconds.

On September 20th it returned to the same circle at 6 hours, 59 minutes, 40 seconds.

The difference is 0 hours, 0 minutes, 7 seconds. Finally, the return of Saturn to the same hourly circle on September 19th occurred when our clock showed 23 hours, 56 minutes, 43 seconds. That fits in well with the amount of Saturn's daily movement, which was in retrograde as seen from the Earth, away from a fixed star which was following it in the diurnal revolution, the one indicated by the letter Θ in the constellation of Capricorn as shown by Bayer, which he calls the forerunner of the two on the back. This star was compared with Saturn in the micrometer of the telescope on the night of September 18th when the clock showed 12 hours, 36 minutes, 42 seconds. This really means that it reached the hourly circle at 11 hours 49m minutes and 16 seconds pm., 3 minutes and 44 seconds after the preceding limb of

D Saturn's ring. On the day before it followed that same edge of Saturn's ring 3 minutes 32 seconds later, and two days before 3 minutes 19 seconds later. Therefore Saturn had moved in retrograde in 24 hours away from the star Θ Capricorni by 12 seconds. And since our clock measures one revolution of a fixed star to the hourly circle, or a daily turning of the celestial sphere, as

23 hours, 56 minutes, 56 seconds, one revolution of Saturn to that same circle was measured as 23 hours, 56 minutes, 44 seconds.

One revolution of Venus to our meridian or to another hourly circle was completed on September 19th, as we have seen from the observation A on our clock, in 23 hours, 59 minutes, 53 seconds: one revolution of Saturn to the same hourly circle in 23 hours 56 minutes 44 seconds. Therefore the change in R.A. was 3 minutes 9 seconds, by which Venus approached Saturn on September 19th in the space of 24 hours, or 189 seconds. Dividing this amount by 24 hours gives $7^7/_8$ seconds of time as the hourly amount by which Venus approached Saturn on its parallel.

E On that same day, September 19th, in the first telescope when the clock showed 1 hour, 40 minutes, $8^1/_2$ seconds, the true time pm. being 52 minutes 53 seconds, the preceding, illuminated limb of Venus touched the micrometer thread marking the hourly circle 1 hour 47 minutes east of the meridian. The preceding limb of Saturn reached this micrometer thread and hourly circle when the pointers of the clock showed 8 hours, 1 minute, 22 seconds (that is, 6 hours, 21 minutes, 13 seconds after the preceding limb of Venus), while on September 18th the edge of Saturn's ring had touched the same thread when the clock showed 8 hours, 4 minutes, 38 seconds. Therefore, the difference in right ascension of the preceding limbs of Venus and Saturn at 0 hours, 52 minutes, 53 seconds pm. on September 19th was 6 hours, 21 minutes, 13 seconds.

F The same day in the fourth telescope the preceding limb of Venus reached the hourly circle when the clock showed 6 hours, 8 minutes, 53 seconds, which was really 5 hours, 21 minutes, 32 seconds pm. The preceding limb of Saturn reached the same hourly circle when the clock showed 12 hours, 29 minutes, 34 seconds, while on the previous day it had touched it at 12 hours, 32 minutes, 46 seconds. The difference, therefore, of right ascension of Venus and Saturn at the true hour 5 pm. 21 minutes and 32 seconds on September 19th was 6 hours, 20 minutes and 36 seconds[9] on our clock.

If no parallax of the globe of Venus had occurred from the time of observation E to the time of observation F, an interval of 4 hours, 28 minutes, 44 seconds, the change in right ascension of Venus and Saturn would have been $35^1/_2$ seconds of time. For as we saw after observation D, that change ought to be $7^7/_8$ seconds every hour.

---

[9] *If the argument is the same as in the previous paragraph, then the difference would appear to be 41 seconds rather than the 36 stated. If the "true time" in F were 5 p.m. 21m 37s i.e. 5 sec. later, making the 36 sec. into 41, then the "interval" in the next paragraph, 4h 28m 44s, makes sense. Also there seems to be a $^1/_2$ second missing from the difference in "E" and the $37^1/_2$ seconds of the following paragraph might have to become $32^1/_2$ to be consistent, if that is indeed the way the argument was intended to go.* **P.F.**

But a comparison of observations E and F shows that the observed variation of right ascension was 37½ seconds. There is, therefore, a difference of 2 seconds of time[10] along the arc of the parallel from the plane of the hourly circle 1 hour 47 minutes east of the meridian, or 26 degrees 45 minutes at the Equator, to the plane of the hourly circle 3 hours 41 minutes west of the meridian, or 55 degrees, 15 minutes at the Equator.

The sine of the angle 26° 45' is	45000
The sine of the angle 55° 15' is	82164
The sum of them both is	127164

As the sum 127,164 is to the complete sine 100,000, so is the parallax of 2 seconds of time, or 120 thirds from each arc, to the 97 seconds parallax corresponding to the complete sine, or to the horizontal parallax of Venus, which is in fact 94 thirds of time, or 1 second and 34 thirds. These 1 seconds and 34 thirds subtend an arc of 0 degrees, 0 minutes, 23 seconds, 30 thirds on the daily parallel of Venus.

This applies to the arc of the great circle as follows. Venus was at 19° south declination. The complement of this angle, 71° 0', has a sine of 94,552 in relation to the complete sine.

As the complete sine of 100,000 is to the sine of 71 degrees, namely 94,552, or as 1000 is to 945, so is 23 seconds, 30 thirds to 22 seconds, 12 thirds. This is the horizontal parallax of Venus which was being sought.

The parallax has been found to be 0 degrees, 0 minutes, 24 seconds, 20 thirds when Venus was observed in conjunction with Regulus on July 3rd, 1716. Therefore, there is sufficient agreement between the parallax found by this more laborious method and the simpler and more precise measurement of 1716. Yet I would not have been so bold as to rely upon this second method alone to determine so minute an angle. But it should be regarded as serving a worthwhile purpose in confirming the measurement of the angle made earlier in the more reliable observation. I must point out, however, that at the time of this second observation in 1727 Venus was a little closer to the Earth than in the previous observation of 1716, and thus our observations ought really to have revealed a slightly larger parallax in the second case than in the first. But the discrepancy of two or three seconds in the case of an angle that is so small for our eyes to perceive should be regarded as so unimportant that it can safely be overlooked.

If it is desired to investigate the Sun's parallax on that day from this slightly smaller value for the parallax of Venus, the calculation should proceed as follows.

---

[10] *2 sec. or 3 sec?*   P.F.

On September 19th the Sun as seen from the Earth was in the 26th degree of Virgo, as is obvious from a comparison of its right ascension with that of Sirius as shown in the observations B and C described above, and also from the ephemerides. Therefore the Earth as seen from the Sun was in the 26th degree of Pisces.

Venus, as seen from the Sun, was in the 26th degree of Aquarius, as is plain from its heliocentric movements given after Chapter VIII, Section III, in the eight-yearly Table. As seen from the Earth it was in the 10th degree of Scorpio, as can be discovered from the observations described above between B and C. Therefore in the triangle STV, given above on Page 133, the angle S at the Sun was 30 degrees, the sine of which is 5000. The angle at the Earth T was 44 degrees. The angle at Venus V was 106 degrees, the sine of which is 9612.

As 9612 is to 5000, so 22 seconds, 12 thirds, or 1332 thirds is to 693 thirds, which equals 11 seconds, 33 thirds. This gives the parallax of the Sun for September 19th. The other observation of July 3rd 1716 gave a value of 14 seconds, 18 thirds, a negligible discrepancy in the case of such a small angle.

XI. If it is desired to repeat with greater accuracy these attempts of mine to establish the parallax of Venus in 1716 and 1727, an opportunity will present itself in July 1732 to try both methods. For Regulus will be near on July 3rd and 4th (admittedly not quite so close as in the years 1716 and 1724), while Saturn will be placed in such a position on July 29th and 30th so that both planets will be almost at the same declination.

It will also be possible to attempt a similar experiment more often by comparing Venus with fixed stars of the first magnitude when in the same declination, particularly with Sirius. But observations with Regulus and Spica are to be preferred above all, as Venus can not only be at the same declination as them, but sometimes very near and in close conjunction. Such an observation would have been possible on August 27th 1727, if clouds had not thwarted the plans of all of us who had been prepared to observe simultaneously from Florence, Bononia and Rome. For it was certain that no suspicion could arise amongst us of discrepancies in timing during the very short period when Regulus and Spica were in such close proximity to Venus.

With this fourth attempt to measure the parallax, therefore, I will close my account of the new phenomena never seen by the ancients on the star of Hesperus and Phosphorus, but reserved for astronomers of our generation to claim as their discoveries, as I have attempted to explain in this short work.

CHAPTER VIII

CONCERNING A MORE OPPORTUNE OCCASION, AND MEANS TO BE ADOPTED, TO REPEAT THE OBSERVATIONS OF THE MARKINGS DESCRIBED ON THE PLANET VENUS AND TO ASCERTAIN THE SPINNING MOVEMENT ROUND ITS OWN AXIS, AND THE PARALLELISM OF THIS AXIS.

Summary of the Chapter

I. The need to seek for the right occasion and means of repeating the observations, because of the difficulty of achieving good results except at stated times and under certain conditions as listed.

II. The choice of a telescope which will magnify the angle of vision a hundred times, and the need to wait for a time when Venus is close and at dichotomy as seen from the Earth.

III. The suitable times can be ascertained from the ephemerides, and also from the 8-year table drawn up according to the planet's heliocentric movements which is included.

IV. Also we gather together various references scattered throughout this short work concerning methods of adapting the telescope to show the markings on the planet, which I have described, with a greater degree of accuracy.

I. I do not think I will have provided adequately for the needs of those who wish to confirm and refine my observations with further ones of their own, if I do not seek to make their task easier as far as I can, by giving them the benefit of our considerable experience over that two-year period since we began our detailed study of the planet's phases. For up to the present time all astronomers, even those equipped with excellent telescopes of 100 or even 200 palms, have experienced great difficulty in detecting any markings at all on Venus. Thus Cassini himself, (whose private letter to a friend in 1667, which we publish below along with a letter of the Very Reverend Father de Briga, is the only evidence I have seen of markings observed on Venus) gave no indication whatever to the scientific community, throughout the forty or more years from that time until his death, of any further observations by himself or anyone else of such a frequently-sought phenomenon.

Nor has anyone else during this long span of time, including thirteen more years since the death of the renowned Cassini to the commencement of our first observations, produced any firm evidence of the sighting of markings on that planet. By way of contrast it has fallen to my lot, not only to obtain a clear view of them myself, but also to show them so plainly to others using the same telescope as myself that, after inspecting the shapes I had drawn on my chart, they were able to confirm during their observations that what they saw was similar. I must forestall the difficulties of unknown future observers to prevent futile efforts to locate these same markings, perhaps through ignorance or neglect of certain precautions, which constant practice has shown to be necessary to educate the eyes of spectators to perceive the evidence of these observations.

II. The first essential is to choose a telescope which magnifies the field of vision a hundred times, made by a skilful craftsman. This alone however is not sufficient to reveal the markings at whatever distance the planet may be. Therefore a telescope of this kind should be used only when the planet Venus is situated so close to the Earth as not to exceed 100 or 120 times the Moon's distance from the Earth. For if the aspect of the planet is studied when it is so situated in its orbit as to be nearer to us than 100 times the Moon's distance, then the contours of the 'maria' or markings, as we have called them, are sufficiently well-defined on that globe's surface through a telescope which magnifies the field of vision a hundred times. Now the reason why this amount of magnification is necessary, though given in Chapter II Section X, should be briefly repeated here.

III. To anyone wishing to undertake observations of this kind, I would suggest that before he turns his eyes and telescopes to the markings or 'maria' he wishes to see on the planet Venus, he should try another experiment when the Moon is at First Quarter or a day or two past. He should enlist the help of an artist skilful at drawing but who has not seen, or at least has had no experience of, the arrangement of the 'maria' observed on the lunar disk and published by several astronomers. He should not allow the artist to use the telescope to make his drawing, but should ask him to make a chart of the markings and a representation of whatever he might be able to see on the Moon with the naked eye. Then he should compare the picture of the Moon drawn by the artist with some careful representation of the lunar globe produced by skilful observers, such as for example the one which the Royal Academy of Sciences published with excellent reproduction and great attention to accuracy, or even the ones produced by Hevelius in his Selenographia and others, though their definition is not quite as perfect. He will certainly see how inaccurately portrayed the boundaries even of the larger markings are by the artist without a telescope (for it is not to be expected that any trace of the very small markings can be seen with the naked eye), and the uncertainty regarding their sizes relative to each other and in proportion to the diameter of the globe. Thus he will understand the truth of Plutarch's remark in his short work "Concerning the face in the Moon's orb", that people with poor eyesight do not perceive the larger lunar markings which we, blessed with normal eyesight, see scattered all over and commonly called 'maria'. Moreover, he will draw a second conclusion from the artist's tentative attempts to portray the shapes and boundaries of the maria, namely that the distance that the lunar markings are from us, 60 times the radius of the Earth, so diminishes the power of the light reflected from those lunar areas that we do not receive a clear impression of their contours everywhere, but only a vague idea of the curves and bays by which these areas of 'maria' are bounded.

After thus preparing his mind through this experiment to show the fallibility of our perception of objects so far distant, the observer should now proceed to the investigation of the Venusian markings which are situated more than a hundred times further away even when the planet is at its nearest. He should use a telescope which magnifies 100 or even 120 times. This does not, however, increase the amount or the strength of the light rays which come to us from it. He should realize that he must not expect equal clarity of definition of the boundaries of the markings on Venus in the telescope, as of the lunar ones as seen with the naked eye. He will be content with a rough sketch of the appearance, to show the relative sizes of the individual markings compared to the planet's disk, and to distinguish the larger from the smaller, the longer from the shorter, the round or elliptical one from the triangular, if there is such, and the half-moon shaped or polygonal, the many-sided or irregular.

Above all, however, he should be sure to observe the planet in the part of its orbit where over a sixth of the illuminated hemisphere is visible. For if only a sixth of the hemisphere facing the Sun is turned towards us, that section of the globe is presented at so oblique an angle, because of the curvature of the surface of the sphere, that only a few rays from that area reach our eyes, and the foreshortening and distortion of the markings is greater on that curve because of the laws of perspective. Therefore the observations of the markings can be started when Venus is at dichotomy after its greatest elongation from the Sun at the evening apparition, or before its greatest elongation from the Sun at the morning apparition for about 30 days, and when it is situated in the part of the Ecliptic that descends steeply when it is called Hesperus and sets after the Sun, or when in the part that ascends steeply, when called Phosphorus and rising before the Sun. It is important to select those days for observing when it seems to us watching from Earth to move forward only half a degree or 40 minutes on the Zodiac. For when it becomes stationary, the illuminated part is already at too oblique an angle, observations are useless because of the slenderness of the crescent, and it is even thinner when in retrograde because of the greater foreshortening.

Since we ought to wait for these conditions regarding the planet, therefore, in order to obtain a sight of the markings in a telescope of 100 or at least 80 palms, it will be expedient for everyone to consult the ephemerides accurately computed by reliable authorities, particularly those of the renowned Eustace Manfred, who recently published information right up to the year 1750, to find out when it will be best to attempt these observations during the next 24 years.

The task can be made easier by following my advice that to work out the times for the next 8 years will suffice. For at the end of an 8-year period, at almost the same day of our civil year, measured according to the Julian and Gregorian calendar, Venus and the Sun reappear in the same degree of the Zodiac to us watching from the Earth, where they were 8 years before. If this is shown in an 8-year table (which we have provided for Venus, for it returns at the same time of year, removing the need for calculation) the times of observation will be seen to be almost identical, and the same will apply to the next 8-year period. (See Table IX.)

After selecting the times when Venus is at the required distance from the Earth, so that the movements of the markings can be seen, the observer should adopt the following aids to obtain a clearer view of them.

He should consult the diagrams above outlining the phases of Venus in Tables I, II and III in order to accustom himself to distinguishing more easily individual markings from others nearby when observing.

To prevent any hindrance to a convenient and thorough representation of the planet's globe in the different parts of its orbit round the Sun as seen by us, and the aspect presented to us by its rotation, I advise the observer to get a solid globe constructed with the celidography marked thereon. For this purpose I have had a final Table made, on which I have drawn twelve sections of the globe showing the markings, using the same method employed by the producers of celestial and terrestrial globes. The strips or sections of this scheme of ours fit a globe whose diameter equals 7 twelfths of the Universal Astronomical Foot established by Cassini, which is equivalent to $7^7/_8$ twelfths of the ancient Roman foot, but to $10^1/_3$ twelfths of the Roman palm used in architecture and longer than 7 twelfths of the Royal Parisian foot by about $1/_8$ of its inches. I have expressed this diameter in all the well-known units of measurement although it is shown in Table X, so that a globe can be made on which the strips will fit more perfectly than if the measurements were to be taken from these pages which are printed while wet and alter in size as they set quite considerably, as everyone knows.

IV. The observer should also consider the choice of eyepiece to be used, which in the case of a telescope of 100 palms should have a focal length of no less than 7 unciae of a Roman palm and no more than 10 unciae[1]. With telescopes of different lengths the proportion should be similar. Nor should he forget the point which we made, that the aperture of the object-glass should be 3 or 4 unciae if the telescope is 100 palms and used on Venus at dichotomy; in fact it should only be 3 unciae when the phase is over half and we see more than half a hemisphere lit by the Sun, but it should be 4 unciae at quadrature and in the crescent phase. A similar proportion should be applied to longer telescopes which must have larger object-glasses, for example 6 unciae for those of 200 palms when used on the gibbous phase, that is to say past quadrature, but 8 unciae at quadrature. The wise observer will also make adjustments to allow for the advance of twilight. Finally, during twilight he should choose a time, as far as is possible, when the Sun is at least half an hour below the horizon at the commencement of observations, to prevent the light reflected from the planet's markings making only a weak impression on the eye, when the air is still boiling with the Sun's rays and lit too strongly with the power of its own light. Nor should he wait for Venus to sink to a lower position than 10 degrees above the horizon. For on lower ground a great deal of earthly mists and vapours pollute the air and diminish too much the power of the rays coming from the planet's disk to the telescope, and entering the eye of the spectator. Everything else can be left to experience and the knowledge acquired by the practised observer of the heavens. To go into greater detail in presuming to advise such a person would be seen as the mark of a tedious bore rather than the zeal of a dutiful conscience.

AT ROME

AT THE PREMISES OF GIOVANNI MARIA SALVIONI
Vatican Typographer in the Principal College of Philosophy.

---

[1] *5.1 to 7.3 of our inches for a telescope of 72.8 feet focal length and aperture either 2.2 or 2.9 inches (13 – 18.5 cm for a telescope of 22.2 m f.l. and aperture 5.6 or 7.4 cm).* **P.F.**

Footnote to the penultimate paragraph:

This paragraph is especially noteworthy. The line at the top of Tab. X in the original book measures 19.04 cm (7.496 inches) by travelling microscope. Then taking $7^1/_8$ inches of the Paris foot to equal 7.5 of our inches, 1 Paris foot would correspond to 1.05(4) of our feet, whilst the data in Chapter II suggest a value of 1.05(6). From the same data $7^7/_8$ twelfths of a Roman foot should come to 7.65 of our inches rather than 7.5. More importantly, if $10^1/_2$ twelfths of the Roman palm equal 7.5 Imperial inches, then 1 palm will correspond to 8.7 Imp. inches, in reasonable agreement with the value of 8.74 inches (to 2 decimal places) provided by Dr. Landels (see footnote in Chapter I.) It is therefore quite impossible to argue that Bianchini intended the Minor Palm of 2.9 Imp. inches .

Furthermore, the paper of the book was made from rag fibres or something similar: it is still in excellent condition, having survived 260 years: our present-day wood pulp paper is most unlikely to last for anything approaching that length of time. Bianchini mentions shrinkage of paper after printing: it was customary to wet printing paper before inserting it into the press, as is clearly depicted in some old woodcuts showing printers at work. Had the paper been perfectly uniform in all directions, then one would expect the drying process to shrink it uniformly too, so that the sections would have fitted a globe of smaller diameter and the line giving the diameter would have shrunk also. On the other hand any process in the manufacture of the paper that introduced non-uniformity in one direction would be expected to cause more shrinking in one direction than in the other. This is indeed what is found in Tab. X, so the remarks in the previous paragraph require some modification. For example, if the line had initially been $7^7/_8$ inches of an ancient Roman foot and then shrank by 15 parts in 765, that would have pointed to a 2% shrinkage across the page, though not necessarily the same down the page. From measurement of the sections it was found that the mean 'great circle' through both poles was 61.7 cm whereas the equator was only 60.5 cm. When some photocopies of the sections were glued onto a football of diameter very nearly the same as the 'line', they suggested a long, thin overcoat on a short, portly person, and the discrepancy is again 2% very nearly. The paper on which Tab. X has been printed has closely spaced watermarking across the page, 29 in 10 cm and wider watermarking down the page, 3 cm. apart. It was almost certainly hand-made because of its date, well before the first known papermaking machine. Hand-made paper tended to be somewhat stronger because by suitable movements of the hands during the forming of the paper, the fibres could be oriented in all directions, producing what was known as 'woven' paper.

It was something of a surprise when the book was shown to Mr. D Knott,

Curator of Rare Books at Reading University Library: he was quite definite that the binding was not of Italian origin but from Northern Europe. The book had probably been sold at some Northern European fair and the binding done either by the bookseller or by the purchaser. The marbled paper on boards was not of the contemporary Italian style and rather cheap. The quarter leather spline with its gold leaf lettering was probably intended to match the other books on a shelf and look suitably expensive: the intending reader would only notice the cheapness of the binding after the book had been taken from the shelf. There were probably expensive leather corners, but these have long since vanished. Examination of the 'paste down' page on the back board revealed a thinner, flimsier paper with watermarks spaced 22 to the inch across the page and a wider spacing down the page exactly 1 inch apart, suggesting that it had been manufactured in Britain.

The diagrams within the text, e.g. the one showing the features on the Moon, were according to Mr. Knott 'mezzotints' and would have had to be impressed on the page after the text had been printed, since the process was quite different from letterpress. In mezzotint a copper plate was roughened uniformly all over and then selectively smoothed in the right places by the correct amount to produce various gradations of shading. It is in fact quite easy to see a rectangular dent in the paper surrounding each of these illustrations, and from the slight misalignment it is obvious that they were added later, possibly by hand.

Lastly it should be mentioned that on the day when the line above Tab, X was measured, the relative humidity was 60% and the absolute value 9 g per kg; there is a fold in the page running across the line and this doubtless caused some small amount of shortening, in spite of efforts made to flatten the paper: as a rough guess the fold may have shortened it by 0.5 to 1 mm. This still leaves 8.7 inches or 22 cm. as a perfectly reliable value for the Palma Romana to two figure accuracy at least. **P.F.**

LETTER

of the Very Reverend Father
MELCHIOR A BRIGA S.J.
Reader in Mathematics at the Florentine College

To the most Illustrious and Reverend Prelate

FRANCESCO BIANCHINI

Concerning the attempts by astronomers to uncover the markings of the Planet Venus.

Translated by Peter Fay

Is it really true? Am I forced to admit that real marks have been seen by mortal eyes on the Mirror of Divine Beauty, the brightest of all bodies, the brilliant star of the morning, *(a)* the planet Venus? Maybe the vulgar and brazen Venus who was gazed at by Paris *(1)* will lack all blemish, but will the heavenly and most chaste Venus *(b)* be not altogether golden in your judgement, you a Prelate in other respects most level headed and wise? Or are the things you call markings really her raiment? For even the clouds are called the sea's garment *(c)* and mistiness is like the cloths of infants *(d)* and they say that the sea itself was given to the Earth in the semblance of a robe *(e)*. What if someone were to come down from outer space and call our seas and lakes marks on the Earth, since the waters reflect poorly the light that they absorb, we should raise no objection to the designation, knowing as we do rightly just how much they may contribute to the adjoining realms in the way of utility and ornament. What? Latium could have boasted its Mirror of Diana: *(2)* why then will not the heavenly Venus have lakes of equal or even greater extent on her planet? Lastly, what we have entitled a favourable and complex likeness, who does not allow us to call it in another context the features of a face? For if the face were always to correspond to a single and altogether uniform appearance, it would not have that beauty which arises from variety in its parts. And indeed as the blackness of the pupil of the eye does not detract from the beauty of the human face but adds to it, so if some part of the face of Venus were to reflect less splendour she will for this very reason be more comely, more lovable and more clearly perceived, if indeed it will prove possible to find out from the marking whether the planet turns on its axis, in what period and in like manner with what inclination to the axis of the ecliptic.

Laying aside the biased opinions which I have till now kept up by way of a joke and weighing up for myself these benefits, when I first tackled the general account of this most lovely planet, I wrote to gentlemen who were my friends, living not only in Italy but also in France, Germany and the most remote empire of the Chinese, (3) that they might consider it worthwhile observing Venus with longer telescopes and not think it too much trouble to reply to me if by chance they were to discover any markings on it. Marks have already been detected in the more remote planets, Mars, Jupiter and Saturn, yes, and even on the Sun itself, even though it is the source of all light in the planetary system. Why then will it not be possible to see them on Venus? Especially when the great Casssini seems to have seen them for himself. Every one of the above wrote back to say that Venus had been observed most diligently by them in response to the entreaties made by me and by my friends using longer and also the best telescopes but no marks could be found on the planet. But you are the one who holds back, worthy prelate, you who have been used to replying most obligingly to my other letters, but to this one letter you had given no reply, with the intention (as you afterwards made plain face to face when you did not disdain to visit me on your journey to Bologna) of excusing the delay, in virtue of your discovery. In the meantime, however, to tell the truth a faint hope remained, though the rest had given up, when I had quite often noticed how little value was the assistance of ordinary telescopes in that investigation.

(a) *Apocalypse XXII 6 "And night shall be no more...."*
(b) *About this double-dealing Venus one should consult Plato in the Symposium etc.*
(c) *Job XXXVIII 9 "When I made a cloud the garment thereof, and*
(d) *Previous reference   wrapped it in a mist as in swaddling bands?"*
(e) *Psalm CIII 6 "The deep like a garment is its clothing...."*

Notes added by the translator are indicated thus: (1)

(1) *Paris, the second son of king Priam of Troy and Hecuba, his wife, had been abandoned by them as an infant, but was brought up by a shepherd. Zeus decided that Paris should settle a dispute between three goddesses as to who was the fairest. Paris judged in favour of Aphrodite, or Venus as the Romans called her, and she promised to reward him with the most beautiful of all women: this proved to be Helen, wife of Menelaus, king of Sparta. The abduction of Helen to Troy sparked off the Trojan War in which the two disappointed goddesses Hera (Juno) and Athena sided with the Greeks, whereas Priam, now reconciled to his son, Paris, relied on the warriors and fortifications of Troy and the goodwill of Aphrodite.*

(2) *Mirror of Diana must refer to some lake in the district of Latium; evidently the still waters reflected the image of the Moon, Diana being the goddess of light associated with the Moon.*

(3) *The Chinese had been noted for their astronomers for many centuries when the missionaries arrived. But one Father Ricci's mathematical knowledge so impressed the Emperor that he gained favour at court: the Jesuit order made the most of their Western knowledge of Astronomy and they were soon running their own observatories, predicting eclipses etc.*

P.F.

There was in addition the complete silence of Cassini after those first doubtful observations at Bologna (I think) and the ineffectual efforts made on this account. For after he was summoned to France by that patron of the sciences, Louis XIV, and presented with longer telescopes, even of 136 Paris feet *(f) (4)* and with them discovered the inner satellites of Saturn *(g)* and the cloud belts of that very pallid and distant planet, and other phenomena difficult to hunt down (h), he failed to find markings on Venus. "Neither from that time", (this was after 1667), "was that mark or a region brighter than it able to be seen" *(i)*. What indeed were the other Academicians to say about that matter when Cassini was silent? In the whole of the History *(4)* of the same Royal Academy of Sciences of France, the best of all by far for reliability in matters of Astronomy, up to the year 1720, the last I have perused, I do not remember reading anything except the conjecture of the well-known Philip de la Hire the elder, who when he saw Venus 'horned' in the year 1700 said "I saw *(k)* in the lower part larger unevenness," (mountains no doubt in the boundary between light and darkness) "than on the Moon, since I have seen other examples of this too. From this it can be surmised that Venus has its markings resembling those of the other planets." Note this mere conjecture of markings, not a clear sighting. Moreover so large an altitude for the Venusian mountains will decrease, if the distance of the Sun from the Earth were to be reduced, as it should: it was this distance that most renowned astronomer, more than the rest of them, increased excessively (Venus will teach us the true distance, as the jewel *(l)* of England, Halley, points out, when it will appear in May 1761 like a spot on the Sun.)

(f)   *See Cassini in Memoirs of the Royal Academy of the Sciences 1705.*

(g)   *These were discovered with a telescope of 100 feet. Du Hamel in History of the Royal Academy 1684. (If these are Paris feet, read 105 British feet.* **P.F.***)*

(h)   *Regarding the mark seen on the fourth satellite of Jupiter, Maraldus in the same memoirs, 1714 p 32 & following, published Amstel. (Amsterdam stands on the banks of the river Amstel.* **P.F.***)*

(i)   *Du Hamel from the Private Papers of the same Academy, dissertation concerning the World & the Heavens II ch. 6 and in the History of the Royal Academy of Sciences, 1670, p103 published Leipzig 1700.*

(k)   *De la Hire in Memoirs 1700 takes all this to mean 'mountains', as does Wolsius, Elements of Astronomy note 463.*

(l)   *Halley in 'Acta Eruditorum' (Journal of Teachers) 1717, p 461. (This was published annually in Leipzig. Halley had pointed this out in 1691; Jeremiah Horrox was the first to witness a transit of Venus in 1639.)*

(4)   *136 Paris feet would be about 143 British feet or 43.5 metres.*

(5)   *Royal Academy of France: 'Histoires' were originally papers and work, 'Memoirs' were accounts of meetings, both published annually.* **P.F.**

To these should be added a third, Huygens, a diligent scrutineer of the stars. When he had watched Venus with a telescope of 60 feet *(Paris feet, probably. P.F.)* when it was nearer the Earth around the time of dichotomy (m), often wondered that it had always appeared brilliant on all sides with even illumination, so that he did not dare assert that any markings had ever been seen by him on the planet, such as are most clearly seen on Jupiter and Mars, even though these planets appear to us smaller. He believed that it was chiefly due to the extreme brilliance of Venus that its markings were least conspicuous. He therefore concluded that it would be best to darken the crystalline lens of the eyepiece with smoke so that quite a large part of the light would be removed. But even with this contrivance it was still not possible to see any lack of uniformity in brightness over the whole face of the planet; either it was because there are no seas on this planet or the waters there reflect the sunlight better than those on our planet or because (as he thought more likely) it is surrounded by a rather dense atmosphere which does not allow the body of the planet to be seen. But the memorable observations of de la Hire conclusively demonstrate the utter falsehood of this, since the unevenness or irregularity of mountains was seen sufficiently clearly by him.

The reason for the difficulty as to why the markings of Venus would be poorly visible is not neglected by Cassini *(n)* in his letter to Petit, where he gives not one but four reasons. Firstly, because they are fainter and less concentrated and since they may be of irregular extent and cover a large part of the visible disc of this planet, they may not appear sufficiently clearly outlined and distinct within their boundaries. Secondly, when Venus is nearest to the Earth (that is around inferior conjunction with the Sun) at what would doubtless have seemed the best time for observing it, it appears near the horizon clouded over by the denser vapours and exhalations of our atmosphere, and with a rather unsteady light, so that its parts only come under inspection in a confused fashion. Thirdly, when it is away from the denser vapours of the horizon, it can only be observed for a brief time, but its motion about its own axis requires a longer space of time in order to be perceptible. Fourthly, when it is closer to the Earth, it shows only a very little part of its disc illuminated, from which the motion about the axis may be discovered, and especially since it is near to the circumference: for the parts near the edge of the disc, sufficiently large otherwise, are scarcely apparent, as

---

(m) The French version of Huygens' Cosmotheoros (Cosmic Spectator) part 2 ch 3. (6)

(6) Cosmotheoros was published after the death of Huygens by his brother. It contains a summary of his final views on Astronomical matters. There is a curious misprint in Bianchini's text, "Cosmocheoros", which is meaningless. **P.F.**

is known from telescopic study, and their motion which of itself is swift, there seems sluggish. For these reasons the same eminent gentleman thought it preferable to observe Venus a little more distant from its inferior conjunction with the Sun and from its nearest point to the Earth, because then it shows a larger phase.

It has been your pleasure, illustrious prelate, to remember all these problems, so that from them one might establish how much the republic of letters *(8)* owes to your remarkable skill; since you were not ignorant of these problems, you straightaway approached the task, which no human has so far been able to accomplish, with remarkable sympathy and courage in response to the entreaties of the unknown Agent. *(9)*

But since I have mentioned the letter of Casini, it will not be out of place to extract from it those things which merit comparison with your observations. I select a French example (although even this has been quoted *(o)* from Cassini's original) since I consider the other versions, whether in Latin or Italian, and the other records of these matters to have been transcribed from the French version, with the times frequently confused or omitted, things which it is wrong for us to mix up or gloss over.

In the first observation Cassini says that he first had his attention drawn to marks on Venus on 14th October 1666 at 5 hours and 45 minutes or so after noon. He saw a certain part shining brighter than the rest near the division (this is how he calls the boundary of light and shadow) not far from the centre on the northern side. At the same time he saw two dark elongated marks towards the west, as in the first diagram, in which (as in the rest) I have taken care to depict the terminator indented, as indeed it appears if viewed with larger telescopes.

Second observation. When during the following days he had sought in vain to see the brighter part, he did not see it again until 28th of April, 1667 one quarter of an hour before sunrise, when Venus was nearly half full.

(n)   See Journal des Scavans 1667, tome 2 p. 257, pub. Amsterdam1676. *(7)*

(7)   *Journal des Scavans (Journal of the Learned) began as a weekly publication in 1665 in France.* **P.F.**

(8)   *Republic of Letters, or 'world of letters'; it hardly refers to any one city or state.*

(9)   *'Unknown Agent' - perhaps he means Bianchini's patron, or the Pope, Clement XI(?) (who for obvious reasons would not wish to be named).* **P.F.**

(o)   *Extract from a letter of Monsieur Cassini, Professor of Astronomy at the University of Bologna, to Monsieur Petit, Superintendant of Fortifications, concerning the discovery that he has made of the movement of the planet Venus about its axis. From June 1667. Volume 2 of the Journal des Scavans is to hand: the citation is on p. 257, published in Amsterdam.*

That part brighter than the rest was visible near the terminator distant from the southern cusp (Cassini speaks like this even though Venus was then not horned but almost half full) a little more than $1/4$ part of the planet's diameter. There was moreover the darker mark, rather elongated, not far from the eastern limb and it was nearer the northern cusp than the southern, as in the second diagram. After sunrise the rather bright part was not quite as

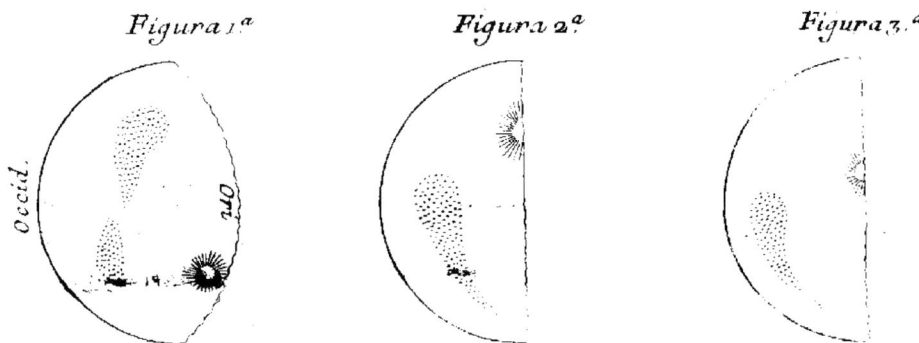

close as hitherto to the southern cusp, from which it was now distant $1/3$ part of the diameter, as in the third diagram *(p)*. Then indeed he rejoiced at having found a clear token of the motion of this planet; but at the time he was surprised that it took place from south to north in the lower hemisphere facing us and from north to south in the upper hemisphere, from which the determination of the size of the motion is facilitated and a name selected for it. Certainly we have no example of motion of this kind, except perhaps in the libration of the Moon.

Third observation. On the next day (i.e. 29th April) towards sunrise this brighter part had scarcely been separated from the division line and was $1/4$ part of the planet's diameter from the southern cusp. When the Sun had risen to 4 degrees above the horizon, it was clearly visible near the division line (I use the author's words) $2/5$ of the diameter away from the southern cusp. Then when the Sun had climbed to 6 degrees 10 minutes it was seen to have jumped across the centre and the terminator (this is the line of illumination that separates the dark part from the illuminated) divided it. Eventually when the Sun had reached 7 degrees it appeared towards the north having moved forward even more, and was divided into two parts by the terminator. From this he infers that there had been some inclination of the motion towards the west.

(p) Ozanam in the Heavenly Sphere, page 80, Figs. 15, 16, 17 shows these same figures, but has added the dark part of Venus, which is least visible and not depicted in the diagrams of Cassini's letter.

Fourth observation. On the 9th of May around sunrise he saw the bright part near the centre of the planet towards the north along with two dark markings which were equally situated between the terminator and the circumference in turn and away from both cusps. Since the weather was calm it was possible for an hour and a half of a quarter to observe its motion, which seemed to take place from the south to the north without any detectible component either eastwards or westwards. In the case of the dark markings and their motion however he noticed so much variation that it could not be put down to optical defect. It is essential here that we consider the last two diagrams.

Fifth observation. On 10th and 13th May before sunrise the brighter part showed up near the centre towards the north.

Sixth and last observation. On 5th and 6th June (of the same year 1667) before sunrise he saw the same part shining with a greater brightness than the rest between the northern cusp and the centre of the planet and he noticed the same irregular variation in the dark markings. Thereafter Venus was moving too far from the Earth and for that reason these phenomena were more difficult to investigate.

After having first put the observations, which we have cited as being a little more obviously distinct, Cassini denies that he could put forward a definite opinion in this matter, as he had once announced concerning the markings of Jupiter and Mars: these may be observed during their opposition with the Sun for nights on end, but the markings of Venus can only be watched for a short time. But if we assume that that part, which had been detected in Venus as outshining the rest, has always been the same one, it is possible to say that its motion fills a period of less than a day, whether it be a motion of rotation or libration, so that in about 23 hours it should return to the same position on the planet: this does not indeed happen without some anomaly or irregularity. Nonetheless even assuming that the brighter part had always been the same, "To say whether this motion occurs through a complete revolution or only by means of a libration, that," he says, "I should not yet venture to maintain, since I have not yet been able to watch the continuity of this motion over a large part of its circle, as in the case of the other planets, and for this very reason it will always be difficult to settle the matter." *(q)*

These final words from so eminent a person, speaking with the greatest circumspection and caution (it is indeed characteristic of great men not to make rash pronouncements), cheered me up wonderfully, since they appeared

---

(q) This footnote gives the original French of the quotation from the letter of Cassini to M. Petit, previously cited.

to leave room for us to investigate whether the markings, that on the final day seemed to have advanced somewhat, had in fact covered that small distance in the space of one day, or whether they had indeed attained the same displacement after completion of one whole revolution. Even if Cassini does indeed incline towards the second hypothesis, that the rotation may be completed in almost 23 hours, yet as is evident from the last words of his letter and from his closest supporter (r) whether that was a whole rotation of the planet or only a libration, he does not dare say with certainty and therefore there is room for the first hypothesis, if better observations more favourable to it were to be brought forward to show that Venus revolved around itself in 23 days. At all events this is what I read in the great European almanacs most recently to see the light of day (s), "After all the observations of the Newtonian school it is found to be established that the planet Venus completes its entire revolution about its own axis in the space of 23 hours". I have not found these observations in Newton's works, nor in Gravesande's Institutions, nor in the works of David Gregory (10), Wishron, Keill etc. But if perhaps they lay hidden in the English Almanacs or Transactions, I do not have these to hand. Nevertheless it is one thing to speak from one's own thoughts, another to do so from observation of Nature. This is foremost.

Secondly I pointed out that in the first of Cassini's observations Venus was Hesperus and seen in the evening, in the rest it was Phosphorus and seen in the morning.

Thirdly, it is obvious from the diagram that the Evening Star in the first observation was ’αμφίκυρτον (11) or gibbous on both sides and this is the phase in which it exhibits itself before it descends to its maximum evening elongation. In the other two diagrams where the Morning Star is depicted διχότομος (dichotomous), the indication is that it was close to its maximum morning elongation. I am therefore surprised that at such a distance from the Earth, such as on the first occasion, and with the Sun already risen, as is stated for the second and third observations, genuine markings could have

---

(r) Du Hamel dissertation II ch. 6 already cited.

(s) From the Gran Giornal di Europa part V for 1667, article 5. "We shall merely add that afterwards all the observations of the School of Newton thus find as established that the planet Venus executes its own complete rotation around its axis in 23 hours."

(10) Gravesande – 1688 to 1742, Dutch scientist accredited with having introduced many ingenious optical and physical devices.

James Gregory – best known for having designed but not built a type of reflecting telescope, was the uncle of the David Gregory mentioned here who taught the Newtonian system in Edinburgh University around the end of the 17th century, before it became established in the two English Universities (Oxford and Cambridge). P.F

(11) Amphikurton - (curved on both sides, not necessarily with the same curvature) - the Latin equivalent gives us 'gibbous'. **P.F.**

been observed on Venus, since those on the Moon seem very subdued and fleeting. For this reason no one, as far as I know, has thought about hunting for markings on Mercury, since the planet's light is swamped by the rays of the Sun.

Fourthly, if we were to say that there have been real markings on the body of the planet, the tilting, that was once barely seen in their motion, can be regarded as merely apparent; but the parallelism of axes is one of the mysteries of real Astronomy: it is remarkable not so much for having escaped the notice of the Greek Astronomers as for not being put to any good use by any of the modern astronomers.

Fifthly, the diagrams of the markings which Cassini shows are a good deal different from those in the globe described by you, learned Prelate, and pointed out to me as bolstering your argument. In this globe I have noticed there are very many markings and these are broader around the planet's equator, almost none around one of the poles of rotation, which ought to shine with a brighter light than other parts, being turned towards us as though in the planet's solstice. When therefore there might be such diversity, it is open to doubt whether the markings of Venus are similar to those which in Jupiter and Mars undergo much transformation: time will tell.

Sixthly. Nonetheless since this is the case, diagrams different from those in the ivory globe *(12)* are exhibited before me, as it were in my imagination. Seeing that whilst Cassini watched they underwent a great change in a short time *(t)*, "Which could not be put down to optical defect"; seeing that after the year 1667 they were sought with longer telescopes throughout nearly 60 years but neither they nor anything like them appeared; seeing that that part that shone more strongly vanished; seeing that they were observed both when the Sun had already risen and when the Evening Star was gibbous far away from the part of its orbit that is nearest the Earth; seeing that in the hemisphere visible to us, that in keeping with its appearance we call the Disc, the motion from south to north is far too different from the motion of the centre (of the Disc) in longitude, I am led to suspect that in truth these phenomena were not on the surface of Venus but in the atmosphere, or more likely in the ether *(13)* intervening between the planet and the eye. Examples of this effect are not lacking.

(t) Above in the fourth observation.

(12) This may have been a reference to a globe of ivory depicting the planet, or to the planet itself. See Chapter IV, Section XI of the text.

(13) The ether – the fifth element of Greek science (the others being earth, air, fire and water) out of which were formed all the 'perfect' bodies of the upper sky. It was believed by Huygens to transmit light waves. **P.F.**

For indeed in 1605 during the lunar eclipse around Easter Michael Maestlin saw in the body of the Moon towards the north a dark mark that occupied nearly a quarter of the Moon's disc. "You might have said," he remarked, "that clouds extended over a large area." *(u)* At Lisbon in 1629 just before the beginning of January a star was seen clinging to the southern cusp of the Moon for up to two days (since on the rest it lay hidden in cloud). However since there was no star reputed to be between the Moon and the Earth, Bettinus (x) came up with the best suggestion: "Some meteorological illusion had deceived the eyes of those gazing upwards". Comets themselves, according to a reasonably probable theory, are produced in no other way than by the accretion of ethereal material into round lumps *(y)* which are set in motion, absorb and reflect light and escape as comets. But not to digress further from our planet, I shall seek an example from Francisco Fontana *(z)*: in 1645 there was seen one and then a second little globe, somewhat blackish or purple, now outside, now beneath the very body of the planet; at that time there was doubt as to whether it might be a satellite of Venus, or a meteor in its atmosphere, or another opaque body between Venus and the observer's eye. There are some who think that there were some *(ϕ)* spots in the glass of Fontana's lenses; but one can not easily suspect this in the case of the learned gentleman, neither should we presume an astronomer to be so ignorant of his speciality that he does not know how to detect this error by rotating the tubes about their own axis, since if it were in the glass it would at once change its position and it would likewise be projected onto other planets viewed with the same tube. *(He means the eyepiece.)* P.F.

What need of more? Cassini himself *(ψ)* appears to have seen one small globe, not blackish but clear, with a telescope of 34 feet *(14)* in 1672 & 1686 (but never at any other time, in spite of his trying his level best) and separated from Venus by $3/5$ths of the length of the planet's diameter and mimicking its phases. Therefore it is unlikely that the phenomenon was in that planet's atmosphere, since no one is readily going to say that it extended out to such a distance. Much less indeed must one suppose it was the same planet's companion that had again made itself clearly visible after so many

---

(u)   See Gassendi, Physics part II book 1 ch.4.
(x)   Bettinus: Apiar. VIII prop. XI.
(y)   Kepler in Physiology of Comets.
(z)   See Riccioli: Almagest Book VIII part 1 ch. 2 p. 485; Gassendi Part2 Physics book 2 ch.2; Tacquet: Astronomy bk. VIII note 33 and following.
(ø)   See the Preface to the Works of Galileo p.17 latest edition.
(ψ)   See David Gregory: Astronomy, book VI, proposition III.
(14) Paris feet, probably.

years of assiduous observation, with the result that at this time it seems rather indecorous for an astronomer to want to crowd the side of Venus with a satellite. It is therefore more credible that the heavenly fluid substance between the eyes of Fontana & Cassini and Venus had been rendered denser at that time so that it could reflect some light *(15)*: this we know has often happened at other times and places in regard to different stars. For this effect it is sufficient that parts of different density should be mixed, as is obvious in the surf becoming whitish, even though it be composed of clear water and transparent air. At those times when the substance has again become rarefied or separated out from the mixture, that whiteness or phenomenon of another colour vanished.

If it so be that on account of the arguments introduced it was permissible to reach the same conclusion about the markings seen by Cassini on our planet, he left intact the prize for the person who can show convincingly that he detected genuine markings on the globe of Venus.

Whatever the outcome may be (and I do not sit here in judgement but only seek evidence) the illustrious gentleman has left this for others to settle whether the new-found motion of that planet is brought about by libration or by a whole revolution and to which part of the sky it is directed.

When, worthy Prelate, you will demonstrate these matters to us, we shall reap the benefit such that no other century will be able to show you as much gratitude. Farewell.

Florence, 7th September 1726.

THE END

---

(15) The same 'ether' (or aether) again; an abrupt change in density at a boundary would correspond to an abrupt change in 'refractive index' (like a bubble in water) and hence cause some reflection of light. (A modern reader might wonder why there is no mention of 'ghost' images - to which some designs of eyepiece are prone, because of internal reflections, especially of bright sources like Venus, Jupiter etc. But if the eyepiece consisted of only a single thin lens, such 'flare spots' would be well out of focus.) Fontana is known to have experimented with eyepiece combinations containing several lenses. **P.F.**

THE AERIAL TELESCOPE

An investigation to find out what users of this type of telescope could actually see.

I realised that the gimbals version would be less prone to troublesome vibrations, whereas the counterweight on a longish arm would be almost impossible to maintain stationary and would be subject to oscillations of long periods in various directions; moreover the observer would need an extremely steady hand to avoid setting up these vibrations. The gimbals, on the other hand, would have a much smaller moment of inertia, since nearly all the mass was concentrated close to the axes and oscillations could be quickly suppressed by the tension in the cord; the cord itself had little mass, so that the velocity of vibrations in it would be high and consequently they would be damped out rapidly, unlike those long tubes which would continue to flex and oscillate for rather long times after each disturbance, whether caused by the observer or by the breeze, on account of their large masses.

There would be one other advantage to the gimbals mounting: the lens would lie on or very close to the point of intersection of the two axes. If in fact its optical centre were to coincide with this point, then any transient oscillation of the lens mounting would merely twist the lens slightly about its optical centre, so that the image of a distant object would remain stationary in the field of view at the eyepiece, provided that the angle of rotation was small. How small? Here was something worth investigating. (It has been said that refractors are less troubled by slight misalignment of the objective lens than are reflectors by misalignment of the primary mirror.) All that would be necessary was for the two perpendicular axes to lie in the same plane and for the lens also to be in this plane, or very close to it: this was hardly asking the impossible.

I resolved therefore to try my hand at making a scaled down version of the aerial telescope to investigate its mechanical stability and handling as well as its optical performance, in so far as that would be feasible. A 0.25 dioptre meniscus spectacle lens (focal length 4 m. diameter 6.0 cm approximately, when mounted) was mounted in a system of 'concentric' tin cans with 'coplanar' bearings and three lengths of waxed cotton thread were fastened to three small bolts spaced 120° apart at the rear of the lens. These threads were similarly fastened to a short plastic tube in which the eyepiece, constructed as closely as possible to the original specifications of Huygens, was packed with foam rubber so that it could be made to slide with a moderate force for focusing: its focal length was 3.3 cm. It worked. Although the bearings were remarkably free, it required only a gentle pull on the threads to maintain the lens pointing at the Moon. The focal tolerance seemed appreciable.

The Moon had been located by turning the objective until it appeared flooded with light and then bringing the eyepiece up to the eye. The threads were 'twanged' to test the stability of the image: it was amazingly stable and, of course, the vibrations died out very rapidly. However, the image quality was poor, although the maria could be recognised. There were three possibilities: the magnification may have been too high or the aperture of the objective too wide or the meniscus shape may have introduced excessive aberrations. Reversing the object lens seemed to make matters worse, and the chromatic aberration was excessive.

I had already decided to make an eyepiece of lower power and this had a focal length of 13.3 cm when completed. Also a plano-convex lens of 0.25 dioptre was ordered. The new eyepiece performed only moderately well with the meniscus objective, sufficiently to allow me to observe a definite crater which turned out to be Bullialdus, but there was an improvement after stopping down the aperture; a stop was also fitted to the eyepiece. But as for the "0.25 dioptre" plano-convex lens, only a blur was seen when this was used: the following day it was checked on a bench and found to be closer to 0.5 dioptre - in fact its focal length when measured carefully was close to 2.0 m! Nevertheless it was not wasted; a 2 m length of black plastic drainpipe was obtained, the lens was fitted and Mizar and Alcor were quite clearly seen with either eyepiece, though surrounded by a great deal of coma.

Meanwhile another plano-convex lens was ordered; it arrived at the shop but was badly chipped and sent back; eventually a replacement did arrive and this time no chances were taken: its focal length was first checked on a bench and found to be close to 5.08 m, which necessitated lengthening the strings.

These two telescopes were tried out whenever a suitable opportunity arose, the Moon when not too high being the clear favourite, although the "drainpipe" could also be turned on bright stars and planets, propped on top of a tall stepladder, since it was just not possible to hold it by hand alone. There were two reasons for the choice of the Moon: (a) it would be easy to tell when the objective was filled with moonlight and then to locate the image with the eyepiece and (b) there was plenty of detail to be seen and any chromatic or spherical aberration would show up very clearly at the lunar limb. It was now increasingly obvious to me that the aerial telescopes of Huygens' time were probably only employed to look at bright objects because of the magnifications mentioned whilst star fields would be next to impossible to locate and the field of view would be very small.

In any case, interest then centred on the Solar System rather than on the much more distant "fixed stars". Evidently one method of aligning the objective and the eyepiece was for a bright lantern to be shone at the objective and the reflected light located in the eyepiece: with the Moon this would hardly be necessary. See "History of the Telescope" by H.C. King pp52 - 64; "Under Newton's Shadow - Astronomical Practices in the Seventeenth Century" by Lesley Murdin p 116 for a historical account of the aerial telescope.

The 2.1 m telescope with the higher power eyepiece, giving a magnification of about 63 times, showed the outline of Saturn on 26 August 1993 unambiguously and almost distinguished the ring: the aberrations did not destroy that much information. Turned to the Moon; with the lower power eyepiece it magnified about 16 times and gave a field of about $1/2°$ and a satisfactory image. Stopped down to 1.1 inch aperture there was no evidence of coloured fringes with the Moon at First Quarter on 19 January 1994, the image was very clear and crisp but not bright enough to upset the eye, and the terminator and brighter parts were very clearly seen: the higher power eyepiece was used.

The aerial telescope, when 4 m long, had been placed on the roof of the garden shed and this was adequate for viewing the Moon on a very few occasions. When the new lens of 5.08 m focal length was fitted, however, the height of the shed roof was quite inadequate. On 25 September 1993 it was strapped to a ladder leaning against the wall of the house; the altitude of the Moon required that the box holding the gimbals should be tilted, the stray light from the Moon that found its way into the eyepiece was a real problem but did not prevent the terminator region from being clearly seen; however, it was necessary to crouch very low and therefore difficult to keep the eyepiece steady for more than a few seconds. Bianchini mentions certain "improvements" to this type of telescope, but does not describe them, apart from mentioning the portable scaffolding which was intended to support the objective mounting and, one presumes, the props for holding the eyepiece steady. I therefore added a means of tilting the box and hanging it from the rungs of a ladder, and put a baffle of cloth around the objective "cell" in an attempt to eliminate stray light: this brought the total weight to about 6 lb. On 21 November 1993 using an extension ladder the box was supported just below the level of the gutter, suitably tilted and revealed what I think was Hipparchus and three other craters on the terminator; the baffle cut off most of the dazzle from stray light which had previously been a problem. The magnification was too high for comfort, some significant chromatic aberration was evident but detail near the terminator was clear even though the rest of the field was confused, especially near the bright limb. Unfortunately the night was far too cold to allow me to continue looking, even though I was able to sit on a low stool. Four nights later when the Moon was again close to the meridian I realised to my surprise that my house was not high enough, being only two-storeyed: a 6 lb mass poised precariously a few feet above the roof tiles did not seem like a bright idea.

This therefore is the present state of the investigation: I await the Moon's being low enough, near the meridian (since the house wall runs nearly north and south), a clear sky and weather not prohibitively chilly. In the meantime I have stopped its aperture down to about 1 inch in anticipation. If I were asked what I thought the observers of the eighteenth century saw with their aerial telescopes, my candid answer must surely be, "not very much and not very often!"

Peter J. Fay
Caversham Park Village, 1st February 1994.

Folded Aerial telescope

Aerial telescope

POSTSCRIPT 16 October 1994

A few nights ago during a recent spell of anticyclonic weather I did at last have the opportunity to view the Moon, just after First Quarter not too high and near the meridian. This time I had the use of a light exending ladder which rendered the whole process very much easier than before. With the lower powered eyepiece the 'Aerial Telescope' would have produced a magnification of about 38 times.

Previously in July I had attempted to do the same but discovered to my surprise that there was excessive vignetting that resisted all my efforts to remove it. There is a saying: "When all else fails, read the instructions". This I did and found that, according to Huygens, the eye ring (the best position for the pupil of the eye) was about 6 cm beyond the eye lens; some simple calculation confirmed this fact. Now if you refer to Tabs. VII and VIII you will notice that the observers' eyes are not close up to the eye lenses, but a few inches away. Since most modern eyepieces have much smaller eye relief, I decided to fit an extension tube with an empty eye lens holder to define where my eye should be: a bench test showed that this gave a much better field of view, and allowed the observer to use it as if it were a modern eyepiece.

With this new arrangement in place, I looked at the Moon. Near the terminator Clavius, Tycho and Eratosthenes were quite easily identified and there was a suggestion of the wall of Copernicus emerging into the light. The field of view was circular but rather small and the image quality can only be described as 'fair'; it was not as good as that produced by the 2m 'drainpipe' telescope, which gave a magnification of about 60 times with the higher powered eyepiece: the rigid tube certainly made for easier control, but then again a tube of '100 palms', or even 22 palms, would have been a very different story. However, my overriding impression was of the extreme difficulty I had in trying to keep the Moon in view, even with the aid of a small stool and a prop to help steady the eye piece. I can only marvel at the skill and patience of Bianchini and his fellow observers who could use these aerial telescopes to produce useful observations such as his diagram of Plato. These experiments certainly made me realise the revolutionary effect that the invention of the achromatic object glass must have had (not to mention the improvements in the manufacture of speculum metal mirrors for reflecting telescopes) upon the whole business of using telescopes for whatever purpose. Far from having my curiosity quenched completely, I am now even more interested in finding out just how these aerial telescopes were used and where, if anywhere, my scaled down copy falls short of Huygens' intentions. But for the present the story must close here.

I wish to acknowledge the kind assistance of Mr A Cripps, Optician, of Caversham in obtaining the lenses that made these experiments possible.

P.F.

TABLE VII

Modi da maneggiare con facilità Canocchiali di qualsisia lunghezza, sip la Terra, che p. il Cielo, inuentati in Roma da Gioseppe Canpani, è praticati in prouare i quattro fabbricati da esso p. l'osseruatorio di S.M.C. che il primo di palmi Romani 105, il sec.º di 130, il terzo di 150, e l'ultimo di 205, dedicati All'Ecc.mo Sig.r di Colbert.

Prima Macchina.

AB	Forma del Canocchiale di figura parallelepipida.
C	Straffone con suo Anello, che abbraccia l'albero, accio si possa girare il Canocchiale.
DE	Asse del Canocchiale.
I	Stromento con rota dentata L, per uoltare, e scortare le corde ciascheduna, delle quali è composta di due, e suo grilletto K, che ferma il rotino
FG	Corda legata in ciascheduno delli quattro lati, e sostentata dalli ponti H con sua scaletta, che si possi alzare e sbassare per render più facilm.te dritto il Canocchiale
QR	Albero di figura Cilindrica, fraposto tra il Canocchiale, et una di dette corde.
R	Girella che si ruolta intorno l'albero, secondo che si gira il Canocchiale sostenuto in aria dal contrapeso S con corda da alzarlo.
T	Caualletto conuste in mezo per alzare, esbassare il Canochiale.

Seconda Macchina.

AB	Forme del Canocchiale di più parti di figura cilindrica
CDEFG	Canale di legno di quattro pezzi CD, DE, EF, FC, sostentato dall asse RQ.
CHI, KHN, MHE, PHS	Corde raddoppiate, che s'interseccano sostentate dalli ponti H nelle cui estremità CG iu è l'instrum.to da scortarle descritto nell'altra macchina.
O	Sostentacolo del Canocchiale, che si possi alzare, et abbassare et metterlo in lineareta.
IL	Varie forme di sostentacoli.

TABLE VIII

Table VIII

Means of handling easily Telescopes of any length for either Terrestrial or Celestial use, invented in Rome by Giuseppe Campani and demonstrated by the four (telescopes) that he made for the observatory of S.M.Cma.[1], the first being 105 Roman palms in length, the second 130, the third 150 and the last 205, dedicated to

The most Excellent Mr di Colbert

First Machine (Oblique telescope)

AB	Shape of the telescope of parallelepipedic construction
C	Strap with associated ring mounted around the tube to allow rotation of the telescope
DE	Axis (of rotation) of the telescope
I	Device with toothed wheel L to loop and guide the ropes, each of which is made of two (ropes or strands); also pawl K which stops the wheel
FG	Ropes tied to each of the four sides (of the telescope) and supported by bridges H, each with its own ratchet, so that they (bridges) be lowered or raised to help with the straightening of the telescope
QR	Cylindrical pole (or mast) between the telescope and one of those ropes
R	Pulley with associated lifting rope which rotates on the pole according to the angle of the telescope whose weight is balanced by the counterweight S
T	Tripod with screw in the middle for raising and lowering the telescope

Second Machine (Horizontal Telescope)

AB	Shape of the telescope assembled from several cylindrical parts
CDEFG	Channel section made of four parts CD, DE, EF, FG supported by the plank (beam) RQ
CHI,KHN	Doubled up ropes which intersect and are supported by MHE,PHS the bridges H at the end of which are guiding devices described for the other machine
D	Support for the telescope which allows it to be raised, lowered or to be set horizontal
IL	(insert top right of figure). Various forms of supports

The above translation of the Italian has been received from Dr. G. Jeronimidis of the Department of Engineering, Reading University.

---

[1] S.M.Cma. This probably means 'His Most Christian Majesty', i.e. Loius XIV, who arranged for the purchase of bigger and better Campani telescopes for Cassini, whom he had persuaded to come to Paris. **P.F.**

TABLE IX

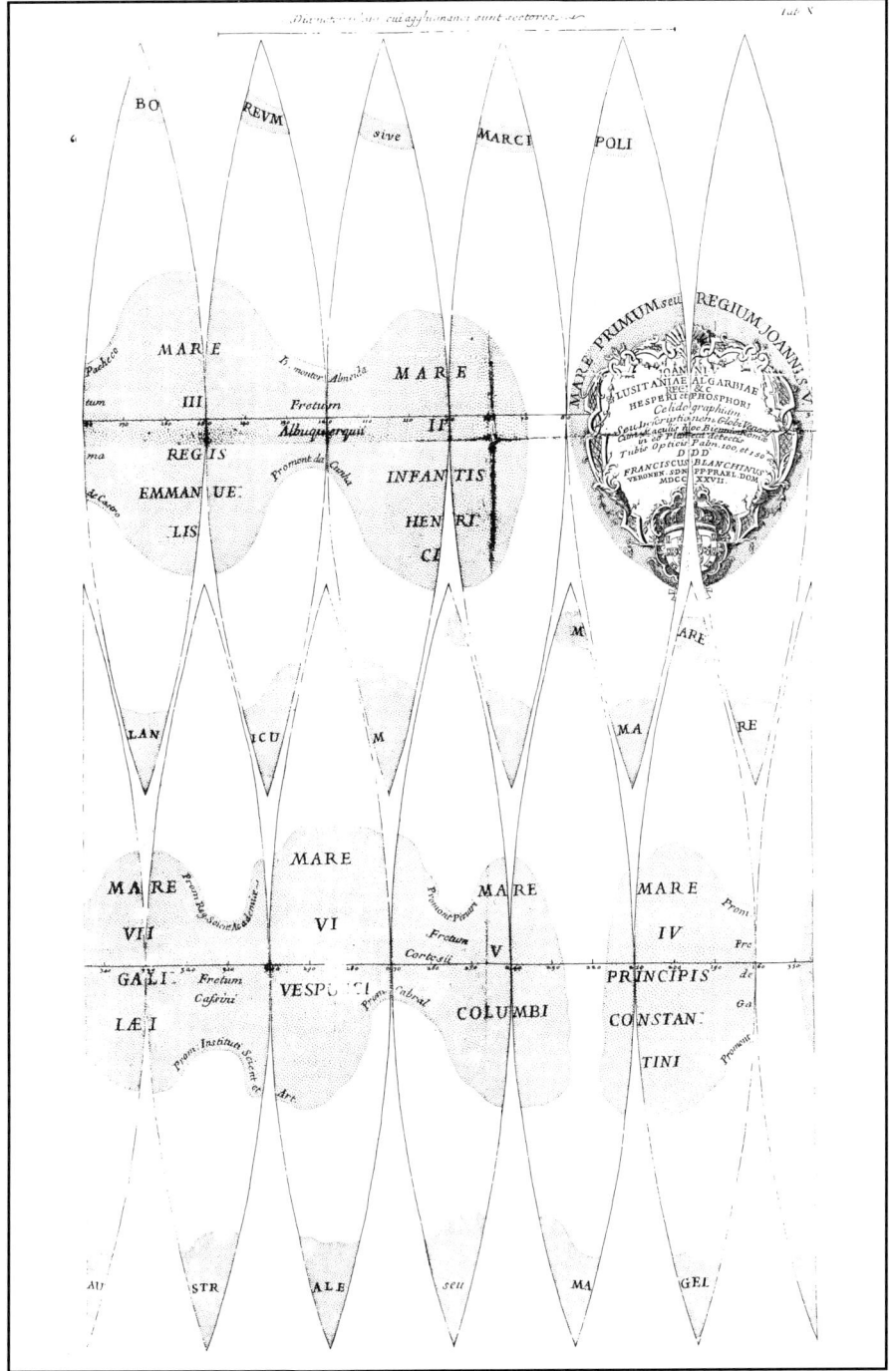

TABLE X This illustration has been reduced to one third size: see next page.

The line at the top of Table X was labelled by Bianchini:

"Diameter of the globe to which the sectors should be glued"

Its length was chosen to be seven twelths of a measurement called the 'universal Astronomical foot' and its length is also given in terms of the other units of measurement found throughout this book, especially the 'Roman palm'.

The line is reproduced again full size in the left margin of this page, because of its significance.

P.F.

ACKNOWLEDGEMENTS As well as those already mentioned, sincere thanks are due to Mr. N.D.Cowell and Mr D. Wilkinson, both of Food Science & Technology, Reading University for their help and encouragement with this work.

P.J.Fay

INDEX OF CHIEF TOPICS

Royal Academy of Sciences established at Paris. 25, 100

Advances in Astronomy made in this century 16

The place found in the ecliptic for the axis of revolution of Venus about itself. 37, 51

The parallelism of the same axis found. 71, 118

The inclination of the same axis to the plane of the ecliptic. 50

The Institute of Sciences & Arts at Bologna. 100

Campani's telescope of 100 palms employed. 21, 27, 45

Observations of Cassini concerning the markings of Venus. 68, 100, 106, 152

*His method applied to the observation of the parallax of Venus by
comparison with the fixed stars.* 124

Celidography or description of the markings detected on the planet Venus. 44

Galileo the first observer of the sickle-shaped phases of Venus. 16

His outstanding scientific arguments. 99

Globe prepared to show the markings of Venus. 103

Inclination of the axis of revolution of Venus to the ecliptic. 50

*The bands of Jupiter, not very different from the markings of Venus,
best observed when it is close to the Earth, not easily seen at its
departure from the Earth.* 42

Outstanding deeds of Italian gentlemen in expeditions to the Indies. 92

*His Excellency, Judice, Duke of Jubenati, is praised in connection with
his patronage, suited to the Sciences.* 27

Lunar markings, which we call seas, cannot be seen by everyone's eyes. 46

*Certain rather recent observations of the Lunar globe made with
telescopes of 100 & 150 palms.* 23

*The favourable progress of the expeditions to the Indies under the
patronage of the kings and princes of Lusitania.* 89

Outstanding achievements of the Lusitanian leaders in these same expeditions. 88/99

Our first observation of the markings of Venus on 9th February 1726. 32

Description of the same. 84

Division into seas, straits & promontories. 86

Celidographic map of the markings of Venus. 81

*The extension of the meridian of Rome through Italy to both the
upper and lower sea.* 27

Names given by us to the markings recently discovered on Venus. 87

Observations of the heavens by recent astronomers. 16

The parallax of Venus accurately determined by means of the fixed stars of first magnitude close to which it was observed. 17, 124

The parallelism of the axis of rotation of Venus about itself determined. 43, 62, 118

The phases of illumination of Venus explained. 36

Recent observations of planets. 16, 17

Planisphere to show the markings of Venus. 102, 103

Path of light recently observed in the Lunar marking Plato. 23

The poles of rotation of Venus about its axis determined 43

Patronage of the Sciences by the Eminent Cardinal of Polignac. 21

Marco Polo of Venice one of the first initiators of journeys to the East. 101

Regulus, or the Lion's Heart, in the neighbourhood of Venus serves to define its parallax. 125

One whole revolution of Venus about its axis is completed in 24 days and about 8 hours. 108, 116

Telescopes of 100 palms or more employed in observations of the markings of Venus. 106

The Terrestrial globe's diameter examined of which the measurement has been found. 26

When first the markings of Venus were observed by us. 32

The spinning of the same about its axis observed. 32, 115

The place in the Zodiac to which the axis of Venus constantly points is about 20 degrees from Leo and Aquarius. 51, 55